LEARN TO **LEAD**

CIVIL AIR PATROL CADET PROGRAMS

LEARN TO **LEAD**

CIVIL AIR PATROL CADET PROGRAMS

"We better be prepared to dominate the skies above the surface of the earth, or be prepared to be buried beneath it."

TOOEY SPAATZ

"To educate a person in mind and not in morals is to educate a menace to society."

THEODORE ROOSEVELT

"The vocation of every man and woman is to serve other people."

LEO TOLSTOY

"I'm just trying to matter, and live a good life, and make work that means something to somebody."

REESE WITHERSPOON

"Two roads diverged in a wood, and I—
I took the one less traveled by,
And that has made all the difference."

ROBERT FROST

GENERAL **MICHAEL E. RYAN**, USAF (RET.)
16TH CHIEF OF STAFF OF THE UNITED STATES AIR FORCE & FORMER CAP CADET

LEARN TO LEAD
Published by Civil Air Patrol
Maxwell Air Force Base, Ala.

CURT LAFOND Series Editor
with Associate Editors
NEIL PROBST & BECCI SUNDHAGEN

LEARN TO **LEAD**

CIVIL AIR PATROL CADET PROGRAMS

CONTENTS

NOTE TO CADETS

This volume is a collection of readings. Unlike the other three volumes of *Learn to Lead*, it is not a textbook that defines key terms, nor does it explain how its teachings are relevant to you in the real world.

Many of the articles here are classics. Their main ideas stand the test of time, though some secondary details might appear a bit dated.

You'll find this volume more challenging than the others. Read each article carefully and apply your own brainpower to identify the author's main ideas, and discern how those principles might be relevant to you in your development as a leader.

12

CHAPTER 12

INTRODUCTION TO STRATEGIC LEADERSHIP

In volume 1, you learned a definition of leadership that applied to individuals and small teams at the tactical level. Volume 2 expanded the concept of leadership to the role of the NCO, educator, creative thinker, motivator, and communicator, still focusing on the tactical and operational levels. This chapter introduces you to various perspectives of leadership at the strategic level. To lead strategically requires careful thought, awareness of systems, and a broad view of your mission. It requires a big picture view, one that focuses on outcomes more than methods, and goals more than tactics.

The chapter starts with a general overview of strategic leadership, provided by Col W. Michael Guillot in "Strategic Leadership: Defining the Challenge." The author will provide you with components, characteristics, and challenges of decision-making at the strategic level, and also provide you with a list of competencies that are essential for strategic leaders.

After you have a clear understanding of the definition of strategic leadership, you will read an example of how grand strategy is implemented at the national level in the White House's "National Security Strategy." In chapter 14 you will trace the development of air power theory over the past century. Look for broad concepts on this topic in this reading. Note that this reading is from the National Security Strategy (NSS) document that was current at the time this textbook went to press. While updates are issued by each administration, the overarching strategic ideas in the NSS are relatively stable.

Moving down to a smaller level, the third article covers the topic of applying systems thinking to problem solving, such as a military force (which comprises one component of a vast national strategy) might use in designing campaigns. In "Leadership and Systems Thinking," Col George E. Reed explains how leaders can apply the art of systems thinking. Echoing the teachings of Peter Senge, Reed urges readers to examine the interrelationships and patterns that present themselves in systems. The idea is to move beyond simple cause-and-effect analysis and find better solutions through more careful examination of system components, behaviors, and relationships.

The fourth article takes you down to a more familiar level, describing how corporations can apply strategic leadership to their decision-making processes. In "Strategic Thinking: Key to Corporate Survival," the authors explore the importance of truly understanding the nature of strategy and strategic planning. They caution that companies that conduct long-range planning incorrectly may actually hinder rather than help their performance.

For a specific example of applying innovative concepts with strategic planning, the final article presents the topic of crowdsourcing. This term refers to the relatively new trend of assigning work to large group of people, who may be highly-skilled amateurs, rather than just a small handful of employees in an organization. The authors of the final article, "Crowdsourcing: What it Means for Innovation," summarize the current state of this concept. As you read the article, you may discover new ways to harness the various talents of a group of people to meet the needs of your project, team, or squadron.

CHAPTER OUTLINE
This chapter's readings are:

Strategic Leadership: Defining the Challenge
Col W. Michael Guillot, "Strategic Leadership: Defining the Challenge," *Air & Space Power Journal* (Winter 2003): 67-75.

National Security Strategy
The White House, "National Security Strategy," (May 2010).

Leadership and Systems Thinking
COL George E. Reed, "Leadership and Systems Thinking," *Defense AT&L* 35, no. 3 (2006): 10-13.

Strategic Thinking: Key to Corporate Survival
Benjamin B. Tregoe and John W. Zimmerman, "Strategic Thinking: Key to Corporate Survival," *Management Review* 68, no. 2 (1979): 8-14.

Crowdsourcing: What it Means for Innovation
Anhai Doan, Raghu Ramarkrishnan, & Alon Y. Halevy, "Crowdsourcing: What it Means for Innovation," *Communications of the ACM* 54, no. 4 (2011): 86-96.

CHAPTER GOALS

1. Comprehend the concept of strategic leadership at the national and organizational level.
2. Summarize the use of systems thinking for strategic planning.
3. Explain how the use of crowd-sourcing technologies can help accomplish team goals.

12.1 Strategic Leadership: Defining the Challenge

By Col W. Michael Guillot, USAF

OBJECTIVES:

1. Define the term "strategic leadership."
2. Identify the four components of the strategic leadership environment, and list factors that belong to each component.
3. Describe four characteristics of consequential decisions.
4. List and define four challenges of strategic leadership.
5. Recall competencies that are essential for leaders who wish to develop strategic leadership skills.

The only thing harder than being a strategic leader is trying to define the entire scope of strategic leadership—a broad, difficult concept. We cannot always define it or describe it in every detail, but we recognize it in action. This type of leadership involves microscopic perceptions and macroscopic expectations. Volumes have been written on the subject, which may in fact contribute to the difficulty of grasping the concept. One finds confusing and sometimes conflicting information on this blended concept that involves the vagaries of strategy and the behavioral art of leadership. Sometimes the methods and models used to explain it are more complicated than the concept and practice of strategic leadership itself. Exercising this kind of leadership is complicated, but understanding it doesn't have to be. Beginning with a definition and characterization of strategic leadership and then exploring components of the strategic environment may prove helpful. Future leaders must also recognize the nature of that environment. Finally, they should also have some familiarity with ways of developing competencies for dealing with the broad, new challenges that are part of leading in the strategic environment.

WHAT IS STRATEGIC LEADERSHIP?

The common usage of the term *strategic* is related to the concept of strategy—simply a plan of action for accomplishing a goal. One finds both broad and narrow senses of the adjective *strategic*. Narrowly, the term denotes operating directly against military or industrial installations of an enemy during the conduct of war with the intent of destroying his military potential.[1] Today, *strategic* is used more often in its broader sense (e.g., strategic planning, decisions, bombing, and even leadership). Thus, we use it to relate something's primary importance or its quintessential aspect—for instance, the most advantageous, complex, difficult, or potentially damaging challenge to a nation, organization, culture, people, place, or object. When we recognize and use *strategic* in this broad sense, we append such meanings as the most important long-range planning, the most complex and profound decisions, and the most advantageous effects from a bombing campaign—as well as leaders with the highest conceptual ability to make decisions.

As mentioned earlier, strategy is a plan whose aim is to link ends, ways, and means. The difficult part involves the thinking required to develop the plan based on uncertain, ambiguous, complex, or volatile knowledge, information, and data. Strategic leadership entails making decisions across different cultures, agencies, agendas, personalities, and desires. It requires the devising of plans that are feasible, desirable, and acceptable to one's organization and partners—whether joint, interagency, or multinational. Strategic leadership demands the ability to make sound, reasoned decisions—specifically, consequential decisions with grave implications. Since the aim of strategy is to link ends, ways, and means, the aim of strategic leadership is to determine the ends, choose the best ways, and apply the most effective means. The strategy is the plan; strategic leadership is the thinking and decision making required to develop and effect the plan. Skills for leading at the strategic level are more complex than those for leading at the tactical and operational levels, with skills blurring at the seams between those levels. In short, one may define strategic leadership as *the ability of an experienced, senior leader who has the wisdom and vision to create and execute plans and make consequential decisions in the volatile, uncertain, complex, and ambiguous strategic environment.*

COMPONENTS OF THE STRATEGIC ENVIRONMENT

What is the strategic-leadership environment? One construct includes four distinct, interrelated parts: the national security, domestic, military, and international environments (fig. 1). Within the strategic environment, strategic leaders must consider many factors and actors. This construct is neither a template nor checklist—nor a recipe for perfection. The framework recognizes the fact that strategic leaders must conceptualize in both the political and military realms. Additionally, it illustrates how the strategic environment is interrelated, complementary, and contradictory. Leaders who make strategic decisions cannot separate the components, especially when they are dealing with the national security environment.

Strategic leaders must recognize and understand the components of the national security environment. The ultimate objectives of all US government personnel are those presented in the national security strategy. The strategy and its objectives shape the decision making of strategic leaders, who must understand the national instruments of power—political, economic, and military.

These instruments provide the means of influence—for example, political persuasion (diplomacy), economic muscle (aid or embargo), or military force (actual or threatened). Within the national security environment, strategic leaders should consider national priorities and opportunities and must know the threats and risks to national security, as well as any underlying assumptions. Understanding this environment poses a major undertaking for strategic leaders. It is also the foundation for understanding the military environment.

Personnel who aspire to be strategic leaders, especially within the Department of Defense, must thoroughly understand military strategy. Two reasons come to mind. First, because the military instrument of power has such great potential for permanent change in the strategic environment, all strategic leaders must recognize its risks and limitations. Second, because military experience among civilian leaders has dwindled over the years and will continue to do so, strategic leaders have a greater responsibility to comprehend policy guidance and clearly understand expected results. Only then can they effectively set military objectives and assess the risks of military operations. Such leaders must develop and evaluate strategic concepts within the military environment and recognize potential threats. Finally, strategic leaders will have to balance capabilities (means) against vulnerabilities and, in doing so, remain aware of the domestic coalition as a major influence.

Since the founding of our nation—indeed, even before the signing of the Constitution—the domestic environment has influenced our leaders. Over the last 200 years, little has changed in this regard; in fact, most people would argue that domestic influence has increased. For instance, strategic leaders today must pay particular attention to the views, positions, and decisions of Congress, whose power and influence pervade many areas within the strategic environment—both foreign and domestic. Congress has the responsibility to provide resources, and we have the responsibility to use them prudently and account for them. This partnership encompasses national and local politics, budget battles for scarce dollars, and cost-risk trade-offs. Strategic leaders cannot ignore either the congressional part of the domestic environment—even though the relationship can sometimes prove difficult—or support from the population. Such support is extremely relevant in democracies and certainly so in the United States. The problem for the strategic leader lies in accurately measuring public support. Accurate or not, senior leaders in a democracy ignore public support at their peril. Actually, because of their power and influence, components of the media make it impossible to ignore domestic issues. Strategic leaders must know how to engage the media since the latter can help shape the strategic environment and help build domestic support. Finally, even though the political will may change, environmental activism will continue to affect the decisions of strategic leaders at every level. Environmental degradation remains a concern for strategic leaders in this country, as do problems in the international environment that call for strategic decisions.

When considering the international environment, strategic leaders should first explore the context—specifically, the history, culture, religion, geography, politics, and foreign security. Who are our allies? Do we have any alliances in place, or do we need to build a coalition? What resources are involved— physical or monetary? Is democracy at stake— creating or defending it? Leaders should also consider threats to the balance of power (BOP) in the environment and the involvement of both official and unofficial organizations. The United Nations may already have

Figure 1

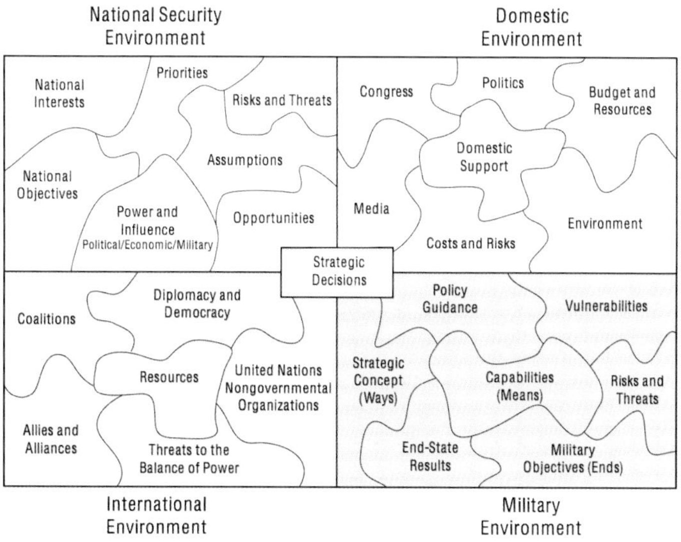

mandates or resolutions that would affect our proposed operations or interests. Nongovernmental organizations may also be willing to help—or perhaps require help. Each of these concerns is legitimate and makes the international environment the most challenging and unfamiliar of them all.

This framework for the components of the strategic environment is simple in design yet complicated in practice. Most US government personnel are intimately familiar with the national security and military environments since they are linked (i.e., military strategy follows directly from national security decisions). But strategic leaders must recognize that the two greatest influences on their decisions come from the domestic and international environments. To lead effectively, they should use what is most familiar and be able to synthesize what influences their strategic decisions.

The four components of the strategic environment present a challenge for strategic leaders. The national security environment, with its many taskmasters, will drive both strategic decisions and military strategy. Leaders will feel great influence from the familiar domestic environment and must have its support for strategic action. Further, strategic leaders can be surprised and their decisions thwarted if they fail to understand the international environment sufficiently. Knowing the disparate components of the strategic environment is the first step in grasping strategic leadership. Understanding the nature of the strategic environment and strategic decisions is the second step.

NATURE OF THE STRATEGIC ENVIRONMENT

The strategic-leadership environment differs from the climate at lower levels of leadership. We should view the nature of this environment both broadly—examining consequential decisions and changes in performance requirements—and narrowly.

CONSEQUENTIAL DECISIONS

By nature, strategic leadership requires consequential decision making. All decisions have consequences, but in the strategic context, they take on a different character—specifically, they are planned, generally long term, costly, and profound.

Consequential decisions occur only at the higher levels within organizations. Generally, decision makers in the top 20 percent of the organization—the people who have ultimate control of resources—plan and execute such decisions. They also think out the implications of their decisions in advance. That is to say, the decision makers analyze and evaluate the possible, probable, and necessary ramifications of a decision beforehand. Some people argue that the sergeant on patrol in Kosovo or the bomber crew over Afghanistan can make strategic decisions in a split second and thus become strategic decision makers. No doubt, armed forces and government officials do make lethal, destructive, and sometimes regrettable decisions. However, these determinations are considered tactical opportunities or, worse, operational blunders rather than planned, consequential decisions. Planning becomes more important when one considers the long-term nature of consequential decisions.

Such decisions require years to play out. Indeed, in most cases strategic decision makers may not be around to witness the actual consequences of the decision, making it all the more essential that they carefully consider all implications before taking action. Clearly, a hasty consequential decision can become very costly.

One may classify these attendant costs as either immediate or mortgaged. For instance, some consequential decisions—such as declaring war or beginning hostilities—can have immediate costs or effects. The cost in lives could become very heavy in a matter of days. World economic costs could mount within weeks while markets collapse within hours. Mortgaged costs of consequential decisions, however, refer to lost opportunities and "sunk" costs. We see such consequences, for example, when organizations commit to huge purchases for weapons systems over a decade-long time frame. Of course in the strategic environment, costs are measured not only in dollars but also in influence (e.g., the costs of supporting one nation over another or the costs of not supporting a particular position). Many times, the decision becomes a matter of sunk costs—gone forever with no chance of recovery. Up to this point, we have considered only the negative effects of costs on consequential decisions. Suffice it to say that many consequential decisions have the aim of decreasing, avoiding, or postponing costs. In fact, some of the least costly consequential decisions turn out to be the most profound (e.g., expanding free-trade agreements and the NATO alliance, reducing the number of nuclear arms, etc.).

Consequential decisions are profound because they have the potential to create great change, lead trends, alter the course of events, make history, and initiate a number of wide-ranging effects. They can change societies and advance new disciplines. Most importantly, an entire organization, a segment of society, a nation, or humanity in general recognizes such decisions as profound.

PERFORMANCE REQUIREMENTS

The stratified systems theory of T. Owen Jacobs and Elliott Jaques classifies the performance requirements for leaders in organizations as direct, general, and strategic (in military parlance: tactical, operational, and strategic, respectively).[2] Distinct elements define the leadership environment within each level. Unmistakable differences among the three levels include complexity, time horizon, and focus.

Most people spend their careers leading at the direct or tactical level (squadron or battalion commander, branch chief, or below). In this environment, the leader interacts directly with the same people every day by maintaining a direct span of control, all the while executing plans, following policies, and consuming resources with a defined goal in mind. The time horizon is very short—normally less than one year. At the direct level of leadership, communications generally occur within the same organization and focus exclusively on the internal audience. Because leaders spend more time at this level than any other, it becomes familiar and comfortable.

Some leaders, however, will mature and move to the general or operational level, where performance requirements begin to change. Direct leadership diminishes as the span of control shrinks. At this level, leaders develop plans, write some policies, and allocate resources among subordinate organizations. The time horizon also increases—to as much as five years. Operational leaders begin to shift the focus of communication and energy outside the organization, recognizing and questioning how the external environment will affect their organizations. Group commanders, brigade commanders, and division chiefs represent this general, analytic level of leadership.

From the perspective of budding strategic leaders, performance requirements for the strategic level change the most and are the least familiar. The power of influence becomes more important than the power of the position. Conceptual ability and communications become essential. Both focus not only on how the external environment will affect the organization, but also—and more importantly—on how the organization can influence that environment. The most challenging of the performance requirements is the time frame for making decisions, which can extend to 20 years and beyond. The leader at this level must think in terms of systems and use integrative thinking—the ability to see linkages and interdependencies within large organizations (or systems) so that decisions in one system will not adversely affect another system.[3] The challenges are great, the stakes are high, and the performance requirements are stringent.

VOLATILITY, UNCERTAINTY, COMPLEXITY, AND AMBIGUITY

Framing the nature of the strategic environment in a broad context helps us understand the magnitude of the challenge. Strategic leaders operate in an environment that demands unique performance requirements for making consequential decisions. If we look more closely at this environment, we discover four characteristics that define the challenge to strategic leadership in a narrow sense: volatility, uncertainty, complexity, and ambiguity.[4]

Now that the world is no longer bipolar, the strategic landscape has become more volatile. Violence erupts in the most unlikely places and for seemingly innocuous reasons. The last few years have given us a glimpse of this volatility: ethnic cleansing in Bosnia and Kosovo, war and terrorism in the Middle East, and terrorism within the United States. The challenge for strategic leaders lies in anticipating volatile scenarios and taking action to avert violence.

In most cases, these leaders will be asked to conduct this action in a landscape of uncertainty—the deceptive characteristic of the strategic environment. They face situations in which the intentions of competitors are not known—perhaps deliberately concealed.[5] At other times, they will even have reservations about the actual meaning of truthful information. Their challenge is to penetrate the fog of uncertainty that hugs the strategic landscape. Comprehending the nature of the strategic environment constitutes the first step toward solving its complexity.

The interdependence of the components in the strategic environment produces complexity—its most challenging characteristic. Integrative thinking is essential to recognizing and predicting the effects of a decision on this "system of systems." If leaders are to anticipate the probable, possible, and necessary implications of the decision, they must develop a broad frame of reference or perspective and think conceptually.

The ambiguous character of the strategic environment stems from different points of view, perspectives, and interpretations of the same event or information. Strategic leaders have to realize that broad perspectives (e.g., using team approaches to solve problems and gain consensus) help eliminate ambiguity and lead to effective strategic decisions.[6]

The nature of the strategic environment is challenging because of the consequences of decisions and unique performance requirements. Although faced with an environment characterized by volatility, uncertainty, complexity, and ambiguity, aspiring strategic leaders can nevertheless learn to master it. Indeed, by acquiring certain skills and competencies, they can transform this environment into something more stable, certain, simple, and clear.

DEVELOPING STRATEGIC LEADERSHIP

If becoming a strategist is the "ends," then leadership is the "ways," and development is the "means." Learning to become a strategic leader requires special preparation in several areas. First, one must understand how such a leader develops—in essence the anatomy of strategic leadership. Second, one should recognize some of the essential competencies a strategic leader must have. Finally, the prospective leader needs to assess his or her current abilities and commit to a development plan.

ANATOMY OF A STRATEGIC LEADER

Development of a strategic leader involves a number of important aspects. First, the most important, indeed foundational, part of this preparation concerns values, ethics, codes, morals, and standards. Second, the path to strategic leadership resembles the building of a pyramid (fig. 2). Shortcuts do not exist, and one can't start at the top—strategic leaders are made, not born. Strategic leaders gradually build wisdom, defined as acquiring experiences over time.[7] One must also remember that certain activities can accelerate these experiences and widen perspectives. Leaders should know that even though some individuals with strategic competency may not become strategic decision makers, they can still influence and contribute to decisions. Additionally, having strategic competency will allow one to fully understand strategic decisions and perspectives.

Figure 2

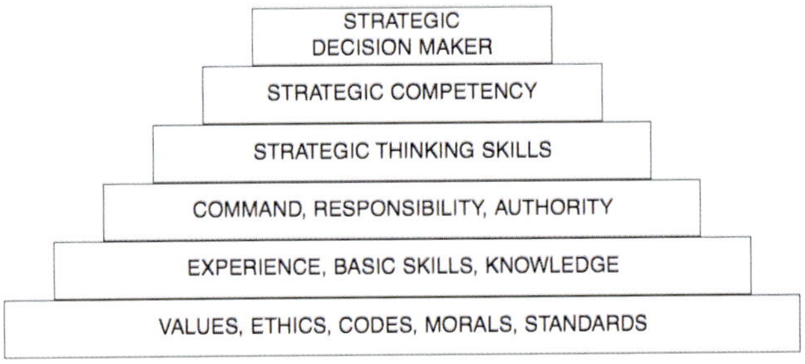

- Strategic leadership begins with organizational values, standards, and ethics—the foundation of our profession.
- Upon this foundation, the officer develops an abstract body of expert knowledge based primarily on experience. Continuing education can influence, expand, and accelerate development.
- Next, the officer is exposed to command responsibility and accountability—a vital phase during which the officer gets his or her first real taste of consequential decision making.
- Further education in strategic-thinking skills enhances the officer's competence. In each case, an officer could have opportunities to exercise strategic competency in support of a strategic leader.
- Ultimately, the officer will participate in strategic decision making and become a strategic leader.

COMPETENCIES

It is difficult to imagine an all-inclusive list of competencies required for strategic leadership. However, some skills seem essential— vision, for instance, which allows the strategic leader to focus on the future and, in fact, build that future. Vision makes leaders proactive in the strategic environment rather than reactive. Furthermore, they should become transformational in order to inspire people toward common goals and shared values; they must anticipate change, lead change, and foster a mind-set of change; they should critically analyze their own thinking to make decisions logically; they should foster an attitude of creativity in their operations and organizations; they must audaciously seek novel ideas and understand how to frame decisions and organize chaos; and they should know how to build effective teams and gain consensus within large organizations. When consensus fails, strategic leaders must negotiate effectively, or they put success at risk. Many times, this kind of success is directly related to the cultural sensitivity and cross-cultural communications ability of the leader. Finally, the strategic leader must assume the role of both teacher and mentor. As Noel Tichy reminds us, great leaders are great teachers. They have a teachable point of view and invest in developing other leaders.[8] The competencies mentioned above form the basis of an education for aspiring strategic leaders.

ASSESSMENT AND DEVELOPMENT

Becoming a strategic leader is a daunting challenge. It starts with taking stock of leadership abilities, conceptual capacity, and interpersonal skills. A thorough self-assessment will help identify strengths and weaknesses. Such assessments can examine personality type, leadership motivation, originality, innovation, tolerance, teamwork, and conceptual ability. These assessments are like the starting point on a map, letting prospective leaders know where they are so they can take the best route to their destination. Completing a detailed self-assessment is also the first step in commitment to the personal and professional development process required to become a strategic leader.

As a follow-up to the self-assessment, aspiring leaders should ask themselves a series of questions: What are my strengths? How can I capitalize on them? Where are my weaknesses? What can I do about them? Where do I want to be in the future? How can I get there? Do I really want to commit to development? The last question is the most difficult one.[9] Those who answer yes are ready to begin the journey toward becoming strategic leaders.

At this point, leader candidates should volunteer for and accept challenging assignments—especially in areas in which they might not have worked before. These could include moving into a different functional area, accepting joint assignments, or working in an interagency environment. Such taskings tend to accelerate experience and broaden perspectives. Furthermore, pursuing a formal course of study at senior service colleges and participating in other education programs would broaden one's knowledge and conceptual ability. Self-learning is also valuable—especially reading. All strategic leaders are voracious readers—and they read outside their normal area of expertise, again, to expand their perspective and increase their conceptual ability. In fact, many of them are experts in a number of unrelated fields. Becoming a "dual expert" helps one think in multiple dimensions.

After committing to some or all of these development activities, potential leaders should reflect on each activity as a way of mining the total benefit and seeking greater meaning. They will also benefit from mentoring other leaders and being mentored themselves. When mentors share their experiences, they help others know and understand them. As Tichy says, sharing experiences or "telling stories" shapes our own attitude, behavior, and point of view.[10] We become the story, and the story guides our lives. Gen Dwight Eisenhower endorsed mentoring when he explained that the best way to become a good decision maker is to be around others who make decisions.[11]

CONCLUSION

The many components of the strategic leadership environment challenge even the best leaders. The monumental consequences of strategic decisions call for individuals with unique performance abilities who can navigate the volatility, uncertainty, complexity, and ambiguity inherent in the nature of those decisions. Aspiring leaders can rise to the challenge by undergoing self-assessment and personal development. Accepting the demands of strategic leadership involves a transition from the art of the familiar to the art of the possible. This is the realm of strategic leadership and the strategic environment.

NOTES

1. Webster's II New Riverside University Dictionary, 1988 ed., s.v. "strategic."

2. T. Owen Jacobs, Strategic Leadership: The Competitive Edge (Fort Lesley J. McNair, Washington, D.C.: Industrial College of the Armed Forces, 2000), 24.

3. US Industrial College of the Armed Forces, chap. 1, "Overview," Strategic Leadership and Decision Making: Preparing Senior Executives for the 21st Century (Washington, D.C.: National Defense University Press, 1997), on-line, Internet, September 2000, available from http://www.ndu.edu/inss/books/books%20-%201999/Strategic%20Leadership%20and%20

Decision-making%20-%20Feb%2099/cont.html.

4. Ibid.

5. Ibid.

6. Ibid.

7. Jacobs, 46.

8. Noel M. Tichy with Eli Cohen, The Leadership Engine: How Winning Companies Build Leaders at Every Level (New York: Harper Business, 1997), 3.

9. US Industrial College of the Armed Forces, chap. 7, "Developing Strategic Leaders," Strategic Leadership and Decision Making.

10. Tichy and Cohen, 77.

11. Edgar F. Puryear Jr., American Generalship: Character Is Everything: The Art of Command (Novato, Calif.: Presidio Press, 2000), 232.

***Col Guillot** is a former cadet from Louisiana and a recipient of the Spaatz Award.*

From: Col W. Michael Guillot, "Strategic Leadership: Defining the Challenge," *Air & Space Power Journal* (Winter 2003): 67-75. Used with permission.

12.2 National Security Strategy

The Administration of President Barack Obama

OBJECTIVES:

6. List the enduring American interests as outlined in the National Security Strategy.
7. Summarize the goals listed in the National Security Strategy related to Diplomacy.
8. Summarize the goals listed in the National Security Strategy related to Strategic Communications.

"More than at any point in human history—the interests of nations and peoples are shared. The religious convictions that we hold in our hearts can forge new bonds among people, or tear us apart. The technology we harness can light the path to peace, or forever darken it. The energy we use can sustain our planet, or destroy it. What happens to the hope of a single child—anywhere—can enrich our world, or impoverish it."

—President Barack Obama,
United Nations General Assembly, September 22, 2009

The United States must renew its leadership in the world by building and cultivating the sources of our strength and influence. Our national security depends upon America's ability to leverage our unique national attributes, just as global security depends upon strong and responsible American leadership. That includes our military might, economic competitiveness, moral leadership, global engagement, and efforts to shape an international system that serves the mutual interests of nations and peoples. For the world has changed at an extraordinary pace, and the United States must adapt to advance our interests and sustain our leadership.

American interests are enduring. They are:

- The security of the United States, its citizens, and U.S. allies and partners;
- A strong, innovative, and growing U.S. economy in an open international economic system that promotes opportunity and prosperity;
- Respect for universal values at home and around the world; and
- An international order advanced by U.S. leadership that promotes peace, security, and opportunity through stronger cooperation to meet global challenges.

Currently, the United States is focused on implementing a responsible transition as we end the war in Iraq, succeeding in Afghanistan, and defeating al-Qa'ida and its terrorist affiliates, while moving our economy from catastrophic recession to lasting recovery. As we confront these crises, our national strategy must take a longer view. We must build a stronger foundation for American leadership and work to better shape the outcomes that are most fundamental to our people in the 21st century.

THE STRATEGIC ENVIRONMENT—THE WORLD AS IT IS

In the two decades since the end of the Cold War, the free flow of information, people, goods and services has accelerated at an unprecedented rate. This interconnection has empowered individuals for good and ill, and challenged state based international institutions that were largely designed in the wake of World War II by policymakers who had different challenges in mind. Nonstate actors can have a dramatic influence on the world around them. Economic growth has alleviated poverty and led to new centers of influence. More nations are asserting themselves regionally and globally. The lives of our citizens—their safety and prosperity—are more bound than ever to events beyond our borders.

Within this environment, the attacks of September 11, 2001, were a transformative event for the United States, demonstrating just how much trends far beyond our shores could directly endanger the personal safety of the American people. The attacks put into sharp focus America's position as the sole global superpower, the dangers of violent extremism, and the simmering conflicts that followed the peaceful conclusion of the Cold War. And they drew a swift and forceful response from the United States and our allies and partners in Afghanistan. This response was followed by our decision to go to war in Iraq, and the ensuing years have seen America's forces, resources, and national security strategy focused on these conflicts.

The United States is now fighting two wars with many thousands of our men and women deployed in harm's way, and hundreds of billions of dollars dedicated to funding these conflicts. In Iraq, we are sup-porting a transition of responsibility to the sovereign Iraqi Government. We are supporting the security and prosperity of our partners in Afghanistan and Pakistan as part of a broader campaign to disrupt, dismantle, and defeat al-Qa'ida and its violent extremist affiliates.

Yet these wars—and our global efforts to successfully counter violent extremism—are only one element of our strategic environment and cannot define America's engagement with the world. Terrorism is one of many threats that are more consequential in a global age. The gravest danger to the American people and global security continues to come from weapons of mass destruction, particularly nuclear weapons. The space and cyberspace capabilities that power our daily lives and military operations are vulnerable to disruption and attack. Dependence upon fossil fuels constrains our options and pollutes our environment. Climate change and pandemic disease threaten the security of regions and the health and safety of the American people. Failing states breed conflict and endanger regional and global security. Global criminal networks foment insecurity abroad and bring people and goods across our own borders that threaten our people.

The global economy is being reshaped by innovation, emerging economies, transition to low-carbon energy, and recovery from a catastrophic recession. The convergence of wealth and living standards among developed and emerging economies holds out the promise of more balanced global growth, but dramatic inequality persists within and among nations. Profound cultural and demographic tensions, rising demand for resources, and rapid urbanization could reshape single countries and entire regions. As the world grows more interconnected, more individuals are gaining awareness of their universal rights and have the capacity to pursue them. Democracies that respect the rights of their people remain successful states and America's most steadfast allies. Yet the advance of democracy and human rights has stalled in many parts of the world.

More actors exert power and influence. Europe is now more united, free, and at peace than ever before. The European Union has deepened its integration. Russia has reemerged in the international arena as a strong voice. China and India—the world's two most populous nations—are becoming more engaged globally. From Latin America to Africa to the Pacific, new and emerging powers hold out opportunities for partnership, even as a handful of states endanger regional and global security by flouting interna-tional norms. International institutions play a critical role in facilitating cooperation, but at times cannot effectively address new threats or seize new opportunities. Meanwhile, individuals, corporations, and civil society play an increasingly important role in shaping events around the world.

The United States retains the strengths that have enabled our leadership for many decades. Our society is exceptional in its openness, vast diversity, resilience, and engaged citizenry. Our private sector and civil society exhibit enormous ingenuity and innovation, and our workers are capable and dedicated. We have the world's largest economy and most powerful military, strong alliances and a vibrant cultural appeal, and a history of leadership in economic and social development. We continue to be a destination that is sought out by immigrants from around the world, who enrich our society. We have a transparent, accountable democracy and a dynamic and productive populace with deep connections to peoples around the world. And we continue to embrace a set of values that have enabled liberty and opportunity at home and abroad.

Now, the very fluidity within the international system that breeds new challenges must be approached as an opportunity to forge new international cooperation. We must rebalance our long-term priorities so that we successfully move beyond today's wars, and focus our attention and resources on a broader set of countries and challenges. We must seize on the opportunities afforded by the world's interconnection, while responding effectively and comprehensively to its dangers. And we must take advantage of the unparalleled connections that America's Government, private sector, and citizens have around the globe.

THE STRATEGIC APPROACH—THE WORLD WE SEEK

In the past, the United States has thrived when both our nation and our national security policy have adapted to shape change instead of being shaped by it. For instance, as the industrial revolution took hold, America transformed our economy and our role in the world. When the world was confronted by fascism, America prepared itself to win a war and to shape the peace that followed. When the United States encountered an ideological, economic, and military threat from communism, we shaped our practices and institutions at home—and policies abroad—to meet this challenge. Now, we must once again position the United States to champion mutual interests among nations and peoples.

Building Our Foundation

Our national security begins at home. What takes place

within our borders has always been the source of our strength, and this is even truer in an age of interconnection.

First and foremost, we must renew the foundation of America's strength. In the long run, the welfare of the American people will determine America's strength in the world, particularly at a time when our own economy is inextricably linked to the global economy. Our prosperity serves as a wellspring for our power. It pays for our military, underwrites our diplomacy and development efforts, and serves as a leading source of our influence in the world. Moreover, our trade and investment supports millions of American jobs, forges links among countries, spurs global development, and contributes to a stable and peaceful political and economic environment.

Yet even as we have maintained our military advantage, our competitiveness has been set back in recent years. We are recovering from underinvestment in the areas that are central to America's strength. We have not adequately advanced priorities like education, energy, science and technology, and health care—all of which are essential to U.S. competitiveness, long-term prosperity, and strength. Years of rising fiscal and trade deficits will also necessitate hard choices in the years ahead.

That is why we are rebuilding our economy so that it will serve as an engine of opportunity for the American people, and a source of American influence abroad. The United States must ensure that we have the world's best-educated workforce, a private sector that fosters innovation, and citizens and busi-nesses that can access affordable health care to compete in a globalized economy. We must transform the way that we use energy—diversifying supplies, investing in innovation, and deploying clean energy technologies. By doing so, we will enhance energy security, create jobs, and fight climate change.

Rebuilding our economy must include putting ourselves on a fiscally sustainable path. As such, imple-menting our national security strategy will require a disciplined approach to setting priorities and mak-ing tradeoffs among competing programs and activities. Taken together, these efforts will position our nation for success in the global marketplace, while also supporting our national security capacity—the strength of our military, intelligence, diplomacy and development, and the security and resilience of our homeland.

We are now moving beyond traditional distinctions between homeland and national security. National security draws on the strength and resilience of our citizens, communities, and economy. This includes a determination to prevent terrorist attacks against the American people by fully coordinating the actions that we take abroad with the actions and precautions that we take at home. It must also include a com-mitment to building a more secure and resilient nation, while maintaining open flows of goods and people. We will continue to develop the capacity to address the threats and hazards that confront us, while redeveloping our infrastructure to secure our people and work cooperatively with other nations.

America's example is also a critical component of our foundation. The human rights which America has stood for since our founding have enabled our leadership, provided a source of inspiration for peoples around the world, and drawn a clear contrast between the United States and our democratic allies, and those nations and individuals that deny or suppress human rights. Our efforts to live our own values, and uphold the principles of democracy in our own society, underpin our support for the aspirations of the oppressed abroad, who know they can turn to America for leadership based on justice and hope.

Our moral leadership is grounded principally in the power of our example—not through an effort to impose our system on other peoples. Yet over the years, some methods employed in pursuit of our security have compromised our fidelity to the values that we promote, and our leadership on their behalf. This undercuts our ability to support democratic movements abroad, challenge nations that violate international human rights norms, and apply our broader leadership for good in the world. That is why we will lead on behalf of our values by living them. Our struggle to stay true to our values and Constitution has always been a lodestar, both to the American people and to those who share our aspiration for human dignity.

Our values have allowed us to draw the best and brightest to our shores, to inspire those who share our cause abroad, and to give us the credibility to stand up to tyranny. America must demonstrate through words and deeds the resilience of our values and Constitution. For if we compromise our values in pur-suit of security, we will undermine both; if we fortify them, we will sustain a key source of our strength and leadership in the world—one that sets us apart from our enemies and our potential competitors.

Pursuing Comprehensive Engagement

Our foundation will support our efforts to engage nations, institutions, and peoples around the world on the basis of mutual interests and mutual respect.

Engagement is the active participation of the United States in relationships beyond our borders. It is, quite simply, the opposite of a self-imposed isolation that denies us the ability to shape outcomes. Indeed, America has never succeeded through isolationism. As the nation that helped to build our international system after World War II and to bring about the globalization that came with the end of the Cold War, we must reengage the world on a comprehensive and sustained basis.

Engagement begins with our closest friends and allies—from Europe to Asia; from North America to the Middle East. These nations share a common history of struggle on behalf of security, prosperity, and democracy. They share common values and a common commitment to international norms that recog-nize both the rights and responsibilities of all sovereign nations. America's national security depends on these vibrant alliances, and we must engage them as active partners in addressing global and regional security priorities and harnessing new opportunities to advance common interests. For instance, we pursue close and regular collaboration with our close allies the United Kingdom, France, and Germany on issues of mutual and global concern.

We will continue to deepen our cooperation with other 21st century centers of influence—including China, India, and Russia—on the basis of mutual interests and mutual respect. We will also pursue diplomacy and development that supports the emergence of new and successful partners, from the Americas to Africa; from the Middle East to Southeast Asia. Our ability to advance constructive cooperation is essential to the security and prosperity of specific regions, and to facilitating global cooperation on issues ranging from violent extremism and nuclear proliferation, to climate change, and global economic instability—issues that challenge all nations, but that no one nation alone can meet.

To adversarial governments, we offer a clear choice: abide by international norms, and achieve the political and economic benefits that come with greater integration with the international community; or refuse to accept this pathway, and bear the consequences of that decision, including greater isolation. Through engagement, we can create opportunities to resolve differences, strengthen the international community's support for our actions, learn about the intentions and nature of closed regimes, and plainly demonstrate to the publics within those nations that their governments are to blame for their isolation.

Successful engagement will depend upon the effective use and integration of different elements of American power. Our diplomacy and development capabilities must help prevent conflict, spur eco-nomic growth, strengthen weak and failing states, lift people out of poverty, combat climate change and epidemic disease, and strengthen institutions of democratic governance. Our military will continue strengthening its capacity to partner with foreign counterparts, train and assist security forces, and pursue military-to-military ties with a broad range of governments. We will continue to foster economic and financial transactions to advance our shared prosperity. And our intelligence and law enforcement agencies must cooperate effectively with foreign governments to anticipate events, respond to crises, and provide safety and security.

Finally, we will pursue engagement among peoples—not just governments—around the world. The United States Government will make a sustained effort to engage civil society and citizens and facilitate increased connections among the American people and peoples around the world—through efforts ranging from public service and educational exchanges, to increased commerce and private sector partnerships. In many instances, these modes of engagement have a powerful and enduring impact beyond our borders, and are a cost-effective way of projecting a positive vision of American leadership. Time and again, we have seen that the best ambassadors for American values and interests are the American people—our businesses, nongovernmental organizations, scientists, athletes, artists, military service members, and students.

Facilitating increased international engagement outside of government will help prepare our country to thrive in a global economy, while building the goodwill and relationships that are invaluable to sus-taining American leadership. It also helps leverage strengths that are unique to America—our diversity and diaspora populations, our openness and creativity, and the values that our people embody in their own lives.

PROMOTING A JUST AND SUSTAINABLE INTERNATIONAL ORDER

Our engagement will underpin a just and sustainable international order—just, because it advances mutual interests, protects the rights of all, and holds accountable those who refuse to meet their responsibilities; sustainable because it is based on broadly shared norms and fosters collective action to address common challenges.

This engagement will pursue an international order that recognizes the rights and responsibilities of all nations. As we did after World War II, we must pursue a rules-based international system that can advance our own interests by serving mutual interests. International institutions must be more effective and representative of the diffusion of influence in the 21st century. Nations must have incentives to behave responsibly, or be isolated when they do not. The test of this international order must be the cooperation it facilitates and the results it generates—the ability of nations to come together to confront common challenges like violent extremism, nuclear proliferation, climate change, and a changing global economy.

That is precisely the reason we should strengthen enforcement of international law and our commitment to engage and modernize international institutions and frameworks. Those nations that refuse to meet their responsibilities will forsake the opportunities that come with international cooperation. Credible and effective alternatives to military action—from sanctions to isolation—must be strong enough to change behavior, just as we must reinforce our alliances and our military capabilities. And if nations challenge or undermine an international order that is based upon rights and responsibilities, they must find themselves isolated.

We succeeded in the post-World War II era by pursuing our interests within multilateral forums like the United Nations—not outside of them. We recognized that institutions that aggregated the national inter-ests of many nations would never be perfect; but we also saw that they were an indispensable vehicle for pooling international resources and enforcing international norms. Indeed, the basis for international cooperation since World War II has been an architecture of international institutions, organizations, regimes, and standards that establishes certain rights and responsibilities for all sovereign nations.

In recent years America's frustration with international institutions has led us at times to engage the United Nations (U.N.) system on an ad hoc basis. But in a world of transnational challenges, the United States will need to invest in strengthening the international system, working from inside interna-tional institutions and frameworks to face their imperfections head on and to mobilize transnational cooperation.

We must be clear-eyed about the factors that have impeded effectiveness in the past. In order for collective action to be mobilized, the polarization that persists across region, race, and religion will need to be replaced by a galvanizing sense of shared interest. Swift and effective international action often turns on the political will of coalitions of countries that comprise regional or international institutions. New and emerging powers who seek greater voice and representation will need to accept greater responsibility for meeting global challenges. When nations breach agreed international norms, the countries who espouse those norms must be convinced to band together to enforce them.

We will expand our support to modernizing institutions and arrangements such as the evolution of the G-8 to the G-20 to reflect the realities of today's international environment. Working with the institutions and the countries that comprise them, we will enhance international capacity to prevent conflict, spur economic growth, improve security, combat climate change, and address the challenges posed by weak and failing states. And we will challenge and assist international institutions and frameworks to reform when they fail to live up to their promise. Strengthening the legitimacy and authority of international law and institutions, especially the U.N., will require a constant struggle to improve performance.

Furthermore, our international order must recognize the increasing influence of individuals in today's world. There must be opportunities for civil society to thrive within nations and to forge connections among them. And there must be opportunities for individuals and the private sector to play a major role in addressing common challenges—whether supporting a nuclear fuel bank, promoting global health, fostering entrepreneurship, or exposing violations of universal rights. In the 21st century, the ability of individuals and nongovernment actors to play a positive role in shaping the international environment represents a distinct opportunity for the United States.

Within this context, we know that an international order where every nation upholds its rights and responsibilities will remain elusive. Force will sometimes be necessary to confront threats. Technology will continue to bring with it new dangers. Poverty and disease will not be completely abolished. Oppression will always be with us. But if we recognize these challenges, embrace America's responsibility to confront them with its partners, and forge new cooperative approaches to get others to join us in overcoming them, then the international order of a globalized age can better advance our interests and the common interests of nations and peoples everywhere.

STRENGTHENING NATIONAL CAPACITY—A WHOLE OF GOVERNMENT APPROACH

To succeed, we must update, balance, and integrate all of the tools of American power and work with our allies and partners to do the same. Our military must maintain its conventional superiority and, as long as nuclear weapons exist, our nuclear deterrent capability, while continuing to enhance its capacity to defeat asymmetric threats, preserve access to the global commons, and strengthen partners. We must invest in diplomacy and development capabilities and institutions in a way that complements and reinforces our global partners. Our intelligence capabilities must continuously evolve to identify and characterize conventional and asymmetric threats and provide timely insight. And we must integrate our approach to homeland security with our broader national security approach.

We are improving the integration of skills and capabilities within our military and civilian institutions, so they complement each other and operate seamlessly. We are also improving coordinated planning and policymaking and must build our capacity in key areas where we fall short. This requires close coopera-tion with Congress and a deliberate and inclusive interagency process, so that we achieve integration of our efforts to implement and monitor operations, policies, and strategies. To initiate this effort, the White House merged the staffs of the National Security Council and Homeland Security Council.

However, work remains to foster coordination across departments and agencies. Key steps include more effectively ensuring alignment of resources with our national security strategy, adapting the education and training of national security professionals to equip them to meet modern challenges, reviewing authorities and mechanisms to implement and coordinate assistance programs, and other policies and programs that strengthen coordination.

- **Defense:** We are strengthening our military to ensure that it can prevail in today's wars; to prevent and deter threats against the United States, its interests, and our allies and partners; and prepare to defend the United States in a wide range of contingencies against state and nonstate actors. We will continue to rebalance our military capabilities to excel at counterterrorism, counterinsurgency, stability operations, and meeting increasingly sophisticated security threats, while ensuring our force is ready to address the full range of military operations. This includes preparing for increasingly sophisticated adversaries, deterring and defeating aggression in anti-access environments, and defending the United States and supporting civil authorities at home. The most valuable component of our national defense is the men and women who make up America's all-volunteer force. They have shown tremendous resilience, adapt-ability, and capacity for innovation, and we will provide our service members with the resources that they need to succeed and rededicate ourselves to providing support and care for wounded warriors, veterans, and military families. We must set the force on a path to sustainable deployment cycles and preserve and enhance the long-term viability of our force through successful recruitment, retention, and recognition of those who serve.

- **Diplomacy:** Diplomacy is as fundamental to our national security as our defense capability. Our diplomats are the first line of engagement, listening to our partners, learning from them, building respect for one another, and seeking common ground. Diplomats, development experts, and others in the United States Government must be able to work side by side to support a common agenda. New skills are needed to foster effective interaction to convene, connect, and mobilize not only other governments and international organizations, but also nonstate actors such as corporations, foundations, nongovern-mental organizations, universities, think tanks, and faith-based organizations, all of whom increasingly have a distinct role to play on both diplomatic and development issues. To accomplish these goals our diplomatic personnel and missions must be expanded at home and abroad to support the increasingly transnational nature of 21st century security challenges. And we must provide the appropriate authorities and mechanisms to implement and coordinate assistance programs and grow the civilian expedi-tionary capacity required to assist governments on a diverse array of issues.

- **Economic:** Our economic institutions are crucial components of our national capacity and our economic instruments are the bedrock of sustainable national growth, prosperity and influence. The Office of Management and Budget, Departments of the Treasury, State, Commerce, Energy, and Agriculture, United States Trade Representative, Federal Reserve Board, and other institutions help manage our currency, trade, foreign investment, deficit, inflation, productivity, and national competitiveness. Remaining a vibrant 21st century economic power also requires close cooperation between and among developed nations and emerging markets because of the interdependent nature of the global economy. America—like other nations—is dependent upon overseas markets to sell its exports and maintain access to scarce commodities and resources. Thus, finding overlapping mutual economic interests with other nations and maintaining those economic relationships are key elements of our national security strategy.

- **Development:** Development is a strategic, economic, and moral imperative. We are focusing on assisting developing countries and their people to manage security threats, reap the benefits of global economic expansion, and set in place accountable and democratic institutions that serve basic human needs. Through an aggressive and affirmative development agenda and commensurate resources, we can strengthen the regional partners we need to help us stop conflicts and counter global criminal networks; build a stable, inclusive global economy with new sources of prosperity; advance democracy and human rights; and ultimately position ourselves to better address key global challenges by growing the ranks of prosperous, capable, and democratic states that can be our partners in the decades ahead. To do this, we are expanding our civilian development capability; engaging with international financial institutions that leverage our resources and advance our objectives; pursuing a development budget that more deliberately reflects our policies and our strategy, not sector earmarks; and ensuring that our policy instruments are aligned in support of development objectives.

- **Homeland Security:** Homeland security traces its roots to traditional and historic functions of government and society, such as civil defense, emergency response, law enforcement, customs, border patrol, and immigration. In the aftermath of 9/11 and the foundation of the Department of Homeland Security, these functions have taken on new organization and urgency. Homeland security, therefore, strives to adapt these traditional functions to confront new threats and evolving hazards. It is not simply about government action alone, but rather about the collective strength of the entire country. Our approach relies on our shared efforts to identify and interdict threats; deny hostile actors the ability to operate within our borders; maintain effective control of our physical borders; safeguard lawful trade and travel into and out of the United States; disrupt and dismantle transnational terrorist, and criminal organiza-tions; and ensure our national resilience in the face of the threat and hazards. Taken together, these efforts must support a homeland that is safe and secure from terrorism and other hazards and in which American interests, aspirations, and way of life can thrive.

- **Intelligence:** Our country's safety and prosperity depend on the quality of the intelligence we collect and the analysis we produce, our ability to evaluate and share this information in a timely manner, and our ability to counter intelligence threats. This is as true for the strategic intelligence that informs executive decisions as it is for intelligence support to homeland security, state, local, and tribal govern-ments, our troops, and critical national missions. We are working to better integrate the Intelligence Community, while also enhancing the capabilities of our Intelligence Community members. We are strengthening our partnerships with foreign intelligence services and sustaining strong ties with our close allies. And we continue to invest in the men and women of the Intelligence Community.

- **Strategic Communications:** Across all of our efforts, effective strategic communications are essential to sustaining global legitimacy and supporting our policy aims. Aligning our actions with our words is a shared responsibility that must be fostered by a culture of communication throughout government. We must also be more effective in our deliberate communication and engagement and do a better job understanding the attitudes, opinions, grievances, and concerns of peoples—not just elites—around the world. Doing so allows us to convey credible, consistent messages and to develop effective plans, while better understanding how our actions will be perceived. We must also use a broad range of meth-ods for communicating with foreign publics, including new media.

- **The American People and the Private Sector:** The ideas, values, energy, creativity, and resilience of our citizens are America's greatest resource. We will support the development of prepared, vigilant, and engaged communities and underscore that our citizens are the heart of a resilient country. And we must tap the ingenuity outside government through strategic partnerships with the private sector, nongovernmental organizations, foundations, and community-based organizations. Such partnerships are critical to U.S. success at home and abroad, and we will support them through enhanced opportunities for engagement, coordination, transparency, and information sharing.

12.3 Leadership and Systems Thinking

By COL George E. Reed, USA

OBJECTIVES:

9. List three steps in the systems thinking approach.
10. Identify barriers to our ability to use systems thinking.

"'For every problem there is a solution that is simple, neat – and wrong.' This maxim has been attributed at various times to Mark Twain, H.L. Mencken, and Peter Drucker as a wake-up call to managers who mistakenly think that making a change in just one part of a complex problem will cure the ails of an entire system. Everyday management thinking too often looks for straightforward cause and effect relationships in problem solving that ignores the effect on, and feedback from, the entire system."

-Ron Zemke,
writing in the February 2011 issue of *Training*

Leaders operate in the realm of bewildering uncertainty and staggering complexity. Today's problems are rarely simple and clear-cut. If they were, they would likely already have been solved by someone else. If not well considered—and sometimes even when they are—today's solutions become tomorrow's problems. Success in the contemporary operating environment requires different ways of thinking about problems and organizations. This article introduces some concepts of systems thinking and suggests that it is a framework that should be understood and applied by leaders at all levels, but especially those within the acquisition community. It is insufficient and often counterproductive for leaders merely to act as good cogs in the machine. Leaders perform a valuable service when they discern that a venerated system or process has outlived its usefulness, or that it is operating as originally designed but against the organization's overall purpose. Sometimes we forget that systems are created by people, based on an idea about what should happen at a given point in time. A wise senior warrant officer referred to this phenomenon as a BOGSAT—a bunch of guys sitting around talking.

SYSTEMS ENDURE

Although times and circumstances may change, systems tend to endure. We seem to be better at creating new systems than changing or eliminating existing ones. Sociologist Robert K. Merton coined the term "goal displacement" to describe what happens when complying with bureaucratic processes becomes the objective rather than focusing on organizational goals and values. When that happens, systems take on a life of their own and seem immune to common sense. Thoughtless application of rules and procedures can stifle innovation, hamper adaptivity, and dash creativity. Wholesale disregard of rules and procedures, however, can be equally disastrous.

When members of an organization feel as though they must constantly fight the system by circumventing established rules and procedures, the result can be cynicism or a poor ethical climate. Because of their experience and position, leaders are invested with the authority to intervene and correct or abandon malfunctioning systems. At the very least, they can advocate for change in a way that those with less positional authority cannot. Leaders at all levels should, therefore, be alert to systems that drive human behavior inimical to organizational effectiveness. It is arguable that military organizations placing a premium on tradition and standardization are predisposed to goal displacement. We need leaders, therefore, who can see both the parts and the big picture; to this end some of the concepts of systems thinking are useful.

The Department of Defense is a large and complex social system with many interrelated parts. As with any system of this type, when changes are made to one part, many others are affected in a cascading and often unpredictable manner. Thus, organizational decisions are fraught with second- and third-order effects that result in unintended consequences. "Fire and forget" approaches are rarely sufficient and are sometimes downright harmful. Extensive planning—combined with even the best of intentions—does not guarantee success. Better prediction is not the answer, nor is it possible. There are so many interactions in complex systems that no individual can be expected to forecast the impact of even small changes that are amplified over time.

GETTING BEYOND THE MACHINE METAPHOR

In her book *Organization Theory: Modern, Symbolic, and Postmodern Perspectives,* Mary Jo Hatch provides an introduction to general systems theory that is useful in thinking about organizations. She makes a point worthy of repeating: The use of lower level models is problematic when applied to higher level systems. Thus, the language of simple machines creates blind spots when used as a metaphor for human or social systems; human systems are infinitely more complex and dynamic. In other words, it can be counterproductive to treat a complex dynamic social system like a simple machine.

Noted management scholar Russell Ackoff puts it another way. He asserts that we are in the process of leaving the machine age that had roots in the Renaissance and came into favor through the industrialization of society. In that era the machine metaphor became the predominant way of looking at organizations. The universe was envisioned by thinkers such as Isaac Newton, as having the characteristics of a big clock. The workings of the clock could be understood through the process of analysis and the analytical method.

Analysis involves taking apart something of interest, trying to understand the behavior of its parts, and then assembling the understanding of the parts into an understanding of the whole. According to Ackoff, "One simple relationship—cause and effect—was sufficient to explain all relationships." Much machine-age thinking remains with us today; however, there are alternatives.

SYSTEMS THINKING

Systems, like the human body, have parts, and the parts affect the performance of the whole. All of the parts are interdependent. The liver interacts with and affects other internal organs—the brain, heart, kidneys, etc. You can study the parts singly, but because of the interactions, it doesn't make much practical sense to stop there. Understanding of the system cannot depend on analysis alone. The key to understanding is, therefore, synthesis. The systems approach is to:

- **Identify a system.** After all, not all things are systems. Some systems are simple and predictable, while others are complex and dynamic. Most human social systems are the latter.

- **Explain the behavior or properties of the whole system.** This focus on the whole is the process of synthesis. Ackoff says that analysis looks into things while synthesis looks out of things.

- **Explain the behavior or properties** of the thing to be explained in terms of the role(s) or function(s) of the whole.

The systems thinker retains focus on the system as a whole, and the analysis in step three (the third bullet) is always in terms of the overall purpose of the system. Borrowing Ackoff's approach and using the example of a contemporary defense issue might help clarify what is admittedly abstract at first glance.

Consider the Institute for Defense Analyses report *Transforming DoD Management: the Systems Approach.* The authors of this study suggested an alternative approach to Service-based readiness reporting, one that considered the entire defense transportation system. One section of the report suggests that knowing the status of equipment, training, and manning of transportation units is helpful but insufficient to determine the readiness of a system that includes elements such as airfields, road networks, ships, and ports. The defense transportation system includes elements of all Services and even some commercial entities. It only makes sense, therefore, to assess readiness of these elements as part of a larger system that has an identifiable purpose—to move personnel and materiel to the right place at the right time. In this example you can clearly see the approach recommended by Ackoff.

THE PROBLEM OF BUSYNESS

Few would disagree, in principle, that senior leaders should see not only the parts, but also the big picture. So why don't we do more of it? One reason is because we are so darned busy. Immersed in the myriad details of daily existence, it is easy to lose sight of the bigger picture. While it may be important to orient on values, goals, and objectives, the urgent often displaces the important. Fighting off the alligators inevitably takes precedence over draining the swamp.

The problem of busyness can be compounded by senior leaders who are overscheduled and uneducated in systems thinking. It seems as though military officers today work excessive hours as a matter of pride. A cursory examination of the calendar of most contemporary officers, especially flag officers, will indicate an abusive pace. Consider as an alternative the example of one of America's greatest soldier-statesmen, Gen. George C. Marshall. Even at the height of World War II, Marshall typically rode a horse in the morning for exercise, came home for lunch and visited with his wife, went to bed early, and regularly took retreats to rejuvenate. To what extent are such pauses for reflection and renewal valued today? Simple cause and effect thinking combined with a culture of busyness can result in decision makers who rapid-fire short-term solutions at long-term problems without taking time to think about the actual impact of those solutions.

A common symptom of this phenomenon can be seen in leaders who unrealistically demand simplicity and certainty in a complex and uncertain environment. The drive for simplicity can lead to the need for excessive assumptions. Few contemporary issues of significance can be understood, much less solved, in a two-page point paper or a PowerPoint® slide. We might also ask whether speed and decisiveness in decision making, so valued at the tactical level, work to the detriment of good decisions at the strategic level. Absent some discipline and techniques to do otherwise, it is very hard to find time for reflection and thoughtful decision making.

Most people expect learning to just happen without their taking the time for thought and reflection, which true learning requires. In the past, with slower communication systems, we often had a few weeks to ponder and rethink a decision. Today we're accustomed to emails, overnight letters, and cell phones, and have come to believe that an immediate response is more important than a thoughtful one.

— Steven Robbins, writing in *Harvard Business School Working Knowledge* in May 2003.

INTERRELATIONSHIPS, NOT THINGS

Peter Senge submits, in *The Fifth Discipline*, that systems thinking provides just the type of discipline and toolset needed to encourage the seeing of "interrelationships rather than things, for seeing patterns of change rather than static 'snapshots.'" Senge argues that this shift of mind is necessary to deal with the complexities of dynamic social systems.

He suggests that we think in terms of feedback loops as a substitute for simple cause and effect relationships. As an example, systems scholar Daniel Aronson suggests that we imagine a farmer who determines that an insect infestation is eating his crop. The conventional approach is to apply a pesticide designed to kill the insect. Our example at this point depicts the lowest level of the thinking hierarchy—reaction. In response to the appearance of insects, the farmer applies a pesticide because he assumes that what has worked in the past will work in this instance. As additional insects appear, the farmer applies more pesticide. While the farmer's goal is to produce a crop, his activity is increasingly consumed by recurring applications of the chemical. He is surely busy, but he may not necessarily be productive. A systems thinker might step back from the problem, take a broader view, and consider what is happening over time.

For example, he might think about whether there are any patterns that appear over weeks or months and attempt to depict what is actually occurring. Recognizing the pattern of a system over time is a higher-order level of thinking. The systems thinker might notice that insect infestation did decrease after applying pesticide, but only for a short time. Insects that were eating the crop were actually controlling a second species of insect not affected by the pesticide. Elimination of the first species resulted in a growth explosion in the second that caused even more damage than the first. The obvious solution caused unintended consequences that worsened the situation.

An accomplished systems thinker would model the above example using a series of feedback and reinforcing loops. The specifics of the modeling technique are less important at this point than the observation that systems thinking tends to see things in terms of loops and patterns aided by constant assessment of what *is* happening, rather than flow charts and reliance on what *should be* happening. At the highest level of thinking, the farmer would try to identify root causes or possible points of intervention suggested by these observations.

THE IMPORTANCE OF CONTINUOUS ASSESSMENT

In *Why Smart Executives Fail*, Sydney Finkelstein examined over 50 of the world's most notorious business failures. His analysis indicated that in almost every case, the failures were not attributable to stupidity or lack of attention. To the contrary, the leaders of well-known corporations such as Samsung Motors, WorldCom, and Enron were exceptionally bright, energetic, and deeply involved in the operation of their businesses. Up to the point of massive corporate failure, they were all extremely successful, and in almost every case, there were some in the organization who vainly raised objections to the course that eventually proved disastrous. In most instances, the executives failed to see or accept what was actually happening. In some cases, they were blinded by their own prior successes; in other cases they inexplicably held tenaciously to a vision, despite plenty of evidence that the chosen strategic direction was ill-advised. The systems thinker's pragmatic focus on determining what is actually happening serves as a preventative to self-delusional wishful thinking. Wishful thinking is no substitute for a realistic appraisal. In the language of systems thinking, the executives were trapped by their own faulty mental models.

The continuous assessment process that is characteristic of systems thinking is essential in a volatile, rapidly changing environment. It takes time and good habits of critical reflection to engage in this kind of learning, both for individuals and organizations.

A systemic approach to failure is more likely to result in effective long-term solutions. Imagine for a moment if the incidents of abuse at Abu Ghraib were chalked up merely to ineffective leadership or just miscreant behavior by some thugs on the night shift. If other factors contributed to the problem, after relieving the chain of command for cause and prosecuting the abusers, the members of the replacement chain of command might have found themselves in an equally untenable situation. While inspired leadership can make a difference under the worst of conditions, we might ask just how heroic we expect our leaders to be on a regular basis. When a system is so obviously stacked against our leaders, there is a moral imperative to change the system.

Systems thinking is no panacea. There is no checklist to work through that will guarantee someone is thinking in a way that will capture the big picture or identify root causes of difficult problems. There are some concepts and approaches embedded in the systems thinking literature, however, that can be very helpful when considering why a situation seems to be immune to intervention, or why a problem thought to be solved has returned with a vengeance. Here are some of the concepts:

• Focus on the purpose for which a system was created over the processes and procedures of the system.

• Simple cause-and-effect relationships are insufficient to understand or explain a complex social system. Patterns over time and feedback loops are a better way to think about the dynamics of complex systems.

• Think in terms of synthesis over analysis; the whole over the parts.

• Busyness and excessive focus on short term gains interferes with our ability to use a systems approach.

• Leaders must see what is actually happening over what they want to see happen.

• Thinking about systems and their dynamics suggests alternative approaches and attunes leaders to important aspects of organizational behavior, especially in military organizations that value tradition and standardization.

ABOUT THE AUTHOR

Reed is the director of command and leadership studies at the United States Army War College in Carlisle, Pa. He has 25 years of experience as a military police officer. He holds a doctorate in public policy analysis and administration.

From: COL George E. Reed, "Leadership and Systems Thinking," *Defense AT&L* 35, no. 3 (2006): 10-13. Used with permission.

12.4 Strategic Thinking: Key to Corporate Survival

By Benjamin B. Tregoe and John W. Zimmerman

OBJECTIVES:

11. Define the term "strategy" as used in this article.
12. Describe the relationship between strategy and operations.
13. Identify reasons why long-range planning impedes strategic thinking.
14. List the advantages of separating strategic thinking from long-range planning.
15. Define the term "driving force" as it relates to long-term strategy.

Most companies face the future unprepared. Though long-range planning has saturated our corporate environment, it does not guarantee success. In our constantly changing environment, the key to corporate survival lies not so much in the quality of our long-range planning as in the clarity of our strategic thinking. To survive and flourish, organizations must face the future knowing *what* they want to be – strategic planning – as well as *how* to get there – long-range planning and operational decision making.

It is our thesis that strategy should provide a picture of the organization as it wants to look in the future. Strategy is vision. It is totally directed at what the organization *should* be rather than *how* the organization will get there. Unfortunately, the word "strategy" has been used rather casually in both management literature and the marketplace. In fact, it has assumed a variety of meanings, some of which confuse the "what" and "how" dimensions.

For example, strategy is sometimes called "strategic planning" and then is used indiscriminately with "long-range planning." Executives talk frequently about a "market strategy" or a "pricing strategy" when they really mean a plan to penetrate a market or a plan to keep prices competitive. Such "strategies" are really major operational decision points that presume an overall corporate or divisional strategy.

While not interested in legislating the meaning of the word strategy, we are interested in avoiding the confusion we have observed. For us, strategy has a very precise meaning, which we define as a *framework that guides those choices that determine the nature and direction of an organization*. These "choices" confront an organization every day. They include choices about an organization's products or services, the geographical markets and customer groups the organization serves, the organization's capabilities of supporting those products and markets, its growth and return, and its allocation of resources.

How these choices are made determines the nature of an organization. If they are made within the context of a strategic framework, the organization's direction is clearly under the control of the managers who develop that framework. If these choices are made in the absence of a strategic framework, you abdicate that control and run the risk of having a direction that is uncoordinated and in the hands of whoever is making these choices.

THE STRATEGY/OPERATIONS RELATIONSHIP

Since strategy sets direction, it must be formulated prior to long-range planning and the day-to-day decision making that flows from such planning. Failure to separate strategy formulation from planning and operations compromises corporate strategic thinking.

The chart below illustrates the relationship between strategy and operations. Clear strategy and effective operations are a winning combination, but with unclear strategy and ineffective operations, you are bound to be a loser. If strategy is clear but operations are ineffective, the result is uncertain – you may still win, but winning depends almost totally on your ability to predict and then be carried by the kindness of external forces such as the economy and competition, forces not generally known for their beneficence. Similarly, if operations are effective but the strategy is unclear, you may survive by being swept forward efficiently – but for how long?

How \ What	Clear Strategy	Unclear Strategy
Effective Operations	Clear strategy and effective operations have equalled success in the past and will in the future	Unclear strategy but effective operations have equalled success in the past, but success doubtful in the future
Ineffective Operations	Clear strategy but ineffective operations have sometimes worked in the past in the short run, but increasing competition makes success doubtful in the future	Unclear strategy and ineffective operations have equalled failure in the past and will in the future

The late W. T. Grant Company is a recent, vivid example of the bottom right quadrant of the chart. It was a loser because it did not have a clear idea of what it should be in the future and had inadequate operations plans. The following commentaries from Business Week attest to Grant's lack of direction:

> Worse yet, early on Grant seemingly could not make up its mind what kind of store it was. "There was a lot of dissension within the company whether we should go the K Mart route or go after the Ward and Penney position," says a former executive. "Ed Staly and Lou Lustenberger were at loggerheads over the issue, with the upshot being we took a position between the two and that consequently stood for nothing."

In addition to its lack of direction, Grant's day-to-day results suffered from ineffective operations:

> From 1963 to 1973 Grant opened 612 stores and expanded 91 others, with the bulk of the increase starting in 1968 under the guidance of president Richard W. Mayer and chairman Edward Staley. "The expansion program placed a great strain on the physical and human capability of the company to cope with the program," says Chairman James G. Kendrick. "These were all large stores – 6 million to 7 million square feet per year – and the expansion of our management organization just did not match the expansion of our stores." Adds a former operations executive: "Our training program couldn't keep up with the explosion of stores, and it didn't take long for the mediocrity to begin to show."

In the upper left quadrant, Sears, Roebuck & Company is typical of a "winner." With a clear image of what it should be in the future, it has also been eminently successful in its operations. While Sears has had its share of trouble recently, over the years it has consistently demonstrated the ability to anticipate needed changes in direction and to organize quickly and efficiently in order to make those changes.

The majority of organizations probably fit in the other two quadrants of the chart. For example, many conglomerates could be placed in the lower left quadrant because they are characterized by well defined growth and financial objectives and ineffective operations. Such organizations tend to see themselves as diverse giants that provide a wide range of products and services. However, the carefully thought-out grand scheme has often been marred by poor operational planning, with resultant over-expansion and inability to manage.

The Swiss watch industry is typical of the upper right quadrant of companies. Superbly efficient at producing and marketing, the industry was overtaken by changes in technology. The Swiss watchmakers' strategy was inadequate to help them anticipate external threats to their survival.

In the United States, strong operations historically have been more important than clear strategic thinking. In the past, many U.S. organizations survived even when they lacked a clear sense of strategic direction. After all, with unlimited resources, skilled labor, and a large, homogeneous market, who needed to think much about what kind of a business they wanted to be in the future?

Now, however, with diminishing resources, world competition, and rising costs, even the most efficient operations may no longer survive the handicap of operating without a clear, strategic direction. Today's company must formulate a clear strategy from which effective operations flow.

LONG-RANGE PLANNING: ROADBLOCK TO STRATEGIC THINKING

Since strategy provides the framework or picture of what the organization wants to be at some future point in time, it must precede and provide the basis for operational planning. Most long-range planning and all short-range planning are operational – they define the "how."

Paradoxically, the real danger to an organization's strategic thinking often comes from its own long-range planning. From our research on strategy, conducted in over 200 major American , Canadian, and European firms, and our strategic-planning consulting with the chief executives of some 75 of these firms, we have seen that primary emphasis on long-range planning impedes strategic thinking. It is ironic that the process on which executives rely most heavily to prepare for the future is doing the most damage, but here is how it happens:

1. *Long-range planning invariably predicts the organization's future by extrapolation from the present.* Projecting from current activities straitjackets the future. Starting with a base of current products and markets makes it difficult to incorporate the new and to eliminate the old in the light of a changing external environment.

2. *Theoreticians who urge the establishment of long-range objectives as a starting point for long-range planning fail to recognize this fact: Most managers do not set objectives that define their future because they lack a process to assist them.* Without practical tools, managers are forced to build their futures on the shaky foundations of the projections instead of on a clear definition of what they want their organizations to be. Where long-range objectives do exist, they are usually set in financial terms. Plans are then developed down the line and are force-fit into the financial constraints imposed by top management. Top executives review these plans and then congratulate

themselves on the realism of their financial objectives, while middle management congratulates itself on its skill in planning to meet those objectives. Planning against objectives that are unconnected to a larger strategy may lead to self-satisfaction; in time, however, it may very well lead to a dead-end future.

3. *Since long-range planning consists of a series of projections about the future, the future picture of the organization can only be a composite of these projections.* Under this approach, the plans companies make determine their direction instead of providing a clear sense of direction determining their plans. Long-range plans are built up from the lowest levels, where information exists to make projections. These projections are additive for the various parts of the organization and, in total, tend to become the recommended plan. But by the time these detailed plans reach the top, there is virtually no opportunity for interjecting fresh insight about the future. In fact, top management's ability to modify these plans, except in minor ways, is practically nil. Flexibility vanishes. The comment of one chief executive immersed in the planning cycle is typical: "By the time we get through with our long-range planning cycle, we are all so engrossed in the precision of our projections that we have lost our ability to question whether they are taking us where we want to go."

4. *Long-range plans invariably tend to be overly optimistic.* This results primarily from the desire of those making the projections at various levels of the organization to do better in their respective areas in the years ahead. By the time this optimism reaches top management, every unit predicts it will do 15 percent better in the years ahead. Such projections tend to become the prevailing corporate wisdom, further restricting the ability of top management to make changes. Any changes that are not purely perfunctory appear arbitrary and capricious to the rest of the organization. Since the allocation of resources is tied to these basically optimistic plans, the persuasiveness of strong personalities and the unrealistic goals they guarantee to reach often determine future resource allocation.

5. *Long-range planning usually begins with assumptions about the environment – the economy, technological change, sociopolitical events, and so on – and the organization's strengths and weaknesses.* Though this information could have great strategic significance, long-range planning tends to utilize such data only as a guide for determining how optimistic or pessimistic to make the long-range product/market projections. This is so because long-range planning is not a process that enables critical information to be used for strategic purposes.

6. *Long-range plans tend to be inflexible (even though they are usually presented in three-ring binders as evidence of their "flexibility").* It takes a tremendous amount of work to project five years ahead; such effort acts as a deterrent to change and transforms most long-range plans into Gothic structures of inflexibility. This inflexibility makes it difficult to react to unanticipated changes int the environment and to adjust plans accordingly. Modification of long-range plans usually occurs only when events reach crisis proportions.

7. *Long-range planning is more short-range than anyone really cares to admit.* To be sure, long-range planning theory suggests that planning should project out five years and then recede back to one year out. But how can this be done in the absence of a framework for looking ahead five years? Without such a structure, the sheer force of necessity leads most managers to reverse the theory and begin by projecting from year one, but beyond that point projects become iffy. Since so much work is involved, the first year usually gets the most thorough analysis. After all, the manager knows he can make changes in the following years; it is only the coming year that cannot be changed – and this year becomes the budget. The shorter the time focus, the more easily a manager is locked in to the constraints of current operations, and the less likely he is to be influenced by information of potential strategic significance. Anyway, most rewards for performance are measured by only first year results.

CHECK YOUR STRATEGY

In summary, strategic thinking is in trouble. Operational long-range planning is no longer adequate to cope with the complexities of today's world.

How is your organization doing? Ask yourself these questions:

- Are your product-market policies and decisions too frequently a reaction to outside influences such as the government, competition, unions, and other outside factors?
- Are acquisition and investment opportunities setting the direction of your company?
- Is the way you are currently organized determining what your company will be doing in the future?
- Do your annual budgets determine what your company will be in the future?
- Do your long-range projections establish the kind of company you will be in the future?
- Do you lack a systematic method to anticipate changes in the environment that may impact your company?
- Do you actually generate assumptions about the environment, but use them for projecting and assessing plans instead of as an input to formulate strategy?
- Is the persuasive manager – the one who is getting

the resources – setting the direction of your company?

- Would different members of your management team paint different pictures of what the company should be in the near and distant future?
- Is your statement of future strategy more helpful for public relations purposes than as a clear guide for future products and markets?

The more of these questions you answered "yes," the more your company's strategy is in trouble. If you answered all "yes," then you can probably hold last rites for strategy in your organization. It is officially dead.

HOW STRATEGIC THINKING CAN SURVIVE

Strategic thinking has long been considered an intellectual nicety; it has provided a patina of respectability to corporate statements built solely on operations considerations. Management attention, however, has been given mainly to operational planning and decisions, for it is here that the "big payoff" could be pursued. In addition, without a process, managers have tended to shy away from the high risks inherent in strategic thinking. They preferred instead to dwell in the lower risk, more secure area of long-range planning. But, in today's world, even the best operations planning and decisions are not enough. We can no longer afford the "security" of avoiding high risk strategic discussions. What, then, can be done?

Strategic thinking must be separated from long-range planning and must precede it. Preaching separation of strategy and long-range planning may appear platitudinous, but most organizations tend to confuse the two. One major corporation, for example, has this patchwork quilt of overall objectives:

> ...to market and produce legitimate products and services at quality levels in their respective markets...to utilize resources fully in order to maximize return on stockholders' investment...to structure the Company and assign responsibility in ways that promote efficiency and incentive, and reward achievement...to provide satisfying, healthful, long-term employment at all levels...to maintain through fluctuating business cycles the confidence of customers, employees, and stockholders...to preserve the integrity of the company in its accounting and reporting procedures, and thereby, the confidence of the investing public.

The first two objectives above say something, however vague, about what the company wants to be in terms of products, markets, and return. But the remaining objectives are operations; they are how-to oriented guidelines for the operation of the business. By masking strategic considerations with operational ones, the above company is headed for an identity crisis as it is pushed and pulled into the future with no clearly defined picture of itself.

Besides making strategic considerations usable, another advantage of separating strategic thinking from operations thinking is that it simplifies the long-range planning process. Strategic thinking and long-range planning in most instances should not cover the same time perspective. A clear, specific statement of strategy covering the next five years generally diminishes the need to project long-range plans over the same time frame. We have found that organizations with clear strategies can put their planning focus on shorter-range plans. Once a strategy is formulated and key areas identified, detailed long-range planning can be limited to these areas.

There is a tendency to feel that because long-range planning covers a longer time span than short-range planning, it is strategic. Conversely, there is a tendency to feel that the short range is not strategic, but operational. Both the operational and the strategic, however, can have either immediate or long-range time significance. Strategy is a function of direction, not time. Operations are a function of *how this direction is achieved*, not time.

Separating strategic from operational thinking also diminishes any controversy over the merits of "top down" versus "bottom up" planning. Both approaches are needed; it is just a matter of where and when. Strategy must be set at the top.

If top management has a unique responsibility, it is to determine the future nature and direction of the organization. Given this strategic framework, the long and short-range operational planning must be done at all levels in the organization where the needed information exists.

If middle and lower levels of management have one unique responsibility, it should be to plan their operations to support the overall direction of the organization.

Once separated from the operational, strategic thinking can survive only if it is clear, specific, and simple. Only then can it provide a framework in which long-range planning and day-to-day decision making can proceed. And only in this way can the executive intelligently assess which products and markets should be emphasized, which should be de-emphasized or abandoned, and what the scope of new products and markets should be. But not many companies have such a framework.

When companies do have a simple statement of corporate strategy, their statements tend to be so general that they are relatively useless as guidelines for specific future product/market choices. Consider this summary statement of corporate strategy:

> Our business is the creation of machines or methods to help find solutions to the increasingly complex problems of businesses, government, science, space exploration, education, medicine, and nearly every area of human endeavor.

Could you establish new market and product priorities based on this?

THE "DRIVING FORCE": KEY TO STRATEGY

The key to developing a simple, clear, and useful statement of strategy lies in the concept of the "driving force." Our research has identified nine strategic areas that impact and influence the nature and direction of any organization. These nine areas can be grouped into three basic categories:

Category	Strategic Areas
1. Products/markets:	Products offered Market needs
2. Capabilities:	Technology Production capability Method of sale Method of distribution Natural resources
3. Results:	Size/growth Return/profit

In every one of the 75 major organizations with which we have worked, we have found that one of the above nine areas can be identified as the *driving force* – the strategic area that is the primary determinant of the organization's products and markets. The driving force also determines the requirements of the organization's other strategic areas.

The following examples, taken from observations of the product and market actions of companies in various industries, further illustrate the concept of the driving force.

1. *Products offered.* The organization with products offered as its driving force will continue to produce products similar to those it has. New products will tend to be very similar to current products, and the organization will seek new markets where there is a need for its existing product line. Its capabilities will be directed toward the support of its basic products. For example, research and engineering would be devoted to product improvements rather than to the development of different kinds of products. The actions of the major automobile companies suggest that their driving force is "products offered."

2. *Market needs.* The organization whose driving force is market needs determine its products or services from needs in the markets or market segments it serves. This organization will constantly look for new and different products to fill these market needs. It will also search for new or emerging needs in these markets. While its capabilities are directed to the support of its current markets and products, it is perfectly willing to acquire very different capabilities to introduce new kinds of products. The actions of major consumer products companies, such as Procter & Gamble, suggest that their driving force is "market needs."

3. *Production capability.* An organization is driven by production capability when it offers products or services that can be performed using its production know-how, equipment, and processes. Looking for economies of scale, it will focus on efficiencies in production, and any new products will utilize the same production know-how, equipment, and processes that produced the original products. The actions of commodity-based companies, such as many of those in the paper industry, suggest that their driving force is (or was) "production capability."

4. *Return/profit.* An organization driven by return/profit will have very specific return/profit targets that may be quite different from its current level of performance. These targets are the basis for developing or acquiring future products and/or markets. Such a driving force will frequently lead this organization into very different and unrelated products or markets as a means of achieving these return/profit objectives over time. The actions of certain conglomerates, such as ITT World Communications, suggest that their driving force is "return/profit."

On first thought many top managers see return/profit as their driving force because profit is equated with survival and is the key measure of continued success. Thus all companies have profit objectives by which to measure operations. Profit, however, is a driving force only if it is the primary determinant of the kinds of future products and markets that characterize an organization. But this is the case in very few companies.

There is no implication in the above examples that the driving force remains fixed. Changes in external events or the desires of top management can change an organization's driving force. A typical pattern of change is from "products offered" to "market needs." For example, this pattern is true for many of the consumer goods and services companies, such as Procter & Gamble, Gillette, Playboy Enterprises, and Merrill Lynch, Pierce, Fenner & Smith.

Another common pattern is to shift from "production capability" to "products offered," a change that has characterized such companies as Kimberly-Clark and International Multifoods (formerly International Milling Company).

Four key reasons explain why the concept of driving force is critical to setting strategy:

- The essential nature of an organization is reflected in its products or services, the markets or customers it services, its capabilities to support these products and markets, and its growth and return. The driving force is the focal point for describing and integrating these key strategic elements.

- Top management discussions to arrive at a driving force bring to the surface issues that must be resolved if an organization is going to arrive at an effective strategy statement. An approach that allows top management to stop short of this will facilitate agreement, but will also result in a general statement of strategy that is no more useful than those previously illustrated.

- Every organization has a momentum that carries it in a certain direction. This momentum is generated by the driving force. Unless the driving force is recognized, attempts to change this direction will be futile. You must know *from what* you are changing. The driving force provides the basic means for thinking about alternative futures and what each might mean in terms of products, markets, capabilities, and return.

- The concept of driving force also has great value in tracking the competition. Since there generally is no way to know the stated strategy of your competitors, assuming they have one, simply observe their actions to determine their driving force and then project what their future courses of action might be.

The rate of change and the complexity of today's world make strategic thinking essential to survival. However, the vehicle that organizations generally have used to cope with the future – long-range planning - is in many ways primarily responsible for stifling their ability to survive and triumph over the challenges ahead. Long-range planning is killing strategic thinking.

Strategic thinking must be separated from and precede long and short-range operational planning. Strategic thinking must result in a statement of strategy that is specific, simple, and clear enough to provide a framework for the determination of future products, markets, capabilities, and return. The *driving force* is the key to developing such a statement.

The chief executives we know voice an increasing sense of urgency about the importance of clear strategic thinking and about their own role in the strategy formulation process. For this urgency to be translated into effective action, top management must devote its most serious and incisive thinking to strategic issues.

ABOUT THE AUTHORS

Benjamin B. Tregoe, Jr., is chairman of the board of directors and chief executive officer of Kepner-Tregoe, Inc., an international management education and organization development firm located in Princeton, New Jersey. Formerly associated with the Rand Corporation, Dr. Tregoe and his colleague, Charles H. Kepner, Ph.D., developed the Kepner-Tregoe Rational Process – the application of cause-effect logic to management activities.

John W. Zimmerman is senior vice-president of the Kepner-Tregoe organization, where his responsibilities include new conceptual research and various aspects of corporate business development. Over the last five years he has specialized with Dr. Tregoe in the development of a program to help chief executive officers clearly determine the future nature and direction of their businesses.

From: Benjamin B. Tregoe and John W. Zimmerman, "Strategic Thinking: Key to Corporate Survival," *Management Review* 68, no. 2 (1979): 8-14. Used with permission.

12.5 Crowdsourcing Systems on the Web

By Anhai Doan, Raghu Ramakrishnan, & Alon Y. Halevy

OBJECTIVES:

16. Identify four challenges that a crowdsourcing system must address.
17. List the dimensions used to classify crowdsourcing systems.
18. Define the roles that humans can play in a crowdsourcing system.
19. Name some common crowdsourcing systems found on the Web.
20. List strategies that crowdsourcing systems can use to recruit and retain users.

Crowdsourcing systems enlist a multitude of humans to help solve a wide variety of problems. Over the past decade, numerous such systems have appeared on the World-Wide Web. Prime examples include Wikipedia, Linux, Yahoo! Answers, Mechanical Turk-based systems, and much effort is being directed toward developing many more.

As is typical for an emerging area, this effort has appeared under many names, including peer production, user-powered systems, user-generated content, collaborative systems, community systems, social systems, social search, social media, collective intelligence, wikinomics, crowd wisdom, smart mobs, mass collaboration, and human computation. The topic has been discussed extensively in books, popular press, and academia.[1,5,15,23,29,35] But this body of work has considered mostly efforts in the physical world.[23,29,30] Some do consider crowdsourcing systems on the Web, but only certain system types[28,33] or challenges (for example, how to evaluate users[12]).

This survey attempts to provide a global picture of crowdsourcing systems on the Web. We define and classify such systems, then describe a broad sample of systems. The sample ranges from relatively simple well-established systems such as reviewing books to complex emerging systems that build structured knowledge bases to systems that "piggyback" onto other popular systems. We discuss fundamental challenges such as how to recruit and evaluate users, and to merge their contributions. Given the space limitation, we do not attempt to be exhaustive. Rather, we sketch only the most important aspects of the global picture, using real-world examples. The goal is to further our collective understanding—both conceptual and practical—of this important emerging topic.

It is also important to note that many crowdsourcing platforms have been built. Examples include Mechanical Turk, Turkit, Mob4hire, uTest, Freelancer, eLance, oDesk, Guru, Topcoder, Trada, 99design, Innocentive, CloudCrowd, and [Crowd Flower]. Using these platforms, we can quickly build crowdsourcing systems in many domains. In this survey, we consider these systems (that is, applications), not the crowdsourcing platforms themselves.

CROWDSOURCING SYSTEMS

Defining crowdsourcing (CS) systems turns out to be surprisingly tricky. Since many view Wikipedia and Linux as well-known CS examples, as a natural starting point, we can say that a CS system enlists a crowd of users to *explicitly* collaborate to build a long-lasting *artifact* that is beneficial to the whole community.

This definition, however, appears too restricted. It excludes, for example, the ESP game,[32] where users *implicitly* collaborate to label images as a side effect while playing the game. ESP clearly benefits from a crowd of users. More importantly, it faces the same human-centric challenges of Wikipedia and Linux, such as how to recruit and evaluate users, and to combine their contributions. Given this, it seems unsatisfactory to consider only explicit collaborations; we ought to allow implicit ones as well.

The definition also excludes, for example, an Amazon's Mechanical Turk-based system that enlists users to find a missing boat in thousands of satellite images.[18] Here, users do not build any artifact, arguably nothing is long lasting, and no community exists either (just users coming together for this particular task). And yet, like ESP, this system clearly benefits from users, and faces similar human-centric challenges. Given this, it ought to be considered a CS system, and the goal of building artifacts ought to be relaxed into the more general goal of solving problems. Indeed, it appears that in principle *any* non-trivial problem *can* benefit from crowdsourcing: we can describe the problem on the Web, solicit user inputs, and examine the inputs to develop a solution. This system may not be practical (and better systems may exist), but it can arguably be considered a primitive CS system.

Consequently, we do not restrict the type of collaboration nor the target problem. Rather, we view CS as a general-purpose problem-solving method. We say that a system is a CS system *if it enlists a crowd of humans to help solve a problem defined by the system owners*, and if in doing so, it addresses the following four fundamental challenges: How to recruit and retain users? What contributions can users make? How to combine user contributions to solve the target problem? How to evaluate users and their contributions?

Not all human-centric systems address these challenges. Consider a system that manages car traffic in Madison, WI. Its goal is to, say, coordinate the behaviors of a crowd of human drivers (that already exist *within* the system) in order to minimize traffic jams. Clearly, this system does not want to recruit more human drivers (in fact, it wants far fewer of them). We call such systems *crowd management* (CM) *systems*. CM techniques (a.k.a., "crowd coordination"[31]) can be relevant to CS contexts. But the two system classes are clearly distinct.

In this survey we focus on CS systems that leverage the Web to solve the four challenges mentioned here (or a significant subset of them). The Web is unique in that it can help recruit a large number of users, enable a high degree of automation, and provide a large set of social software (for example, email, wiki, discussion group, blogging, and tagging) that CS systems can use to manage their users. As such, compared to the physical world, the Web can dramatically improve existing CS systems and give birth to novel system types.

Classifying CS systems. CS systems can be classified along many dimensions. Here, we discuss nine dimensions we consider most important. The two that immediately come to mind are the *nature of collaboration* and *type of target problem*. As discussed previously, collaboration can be explicit or implicit, and the target problem can be any problem defined by the system owners (for example, building temporary or permanent artifacts, executing tasks). The next four dimensions refer respectively to how a CS system solves the four fundamental challenges described earlier: *how to recruit and retain users; what can users do; how to combine their inputs; and how to evaluate them*. Later, we will discuss these challenges and the corresponding dimensions in detail. Here, we discuss the remaining three dimensions: degree of manual effort, role of human users, and standalone versus piggyback architectures.

Degree of manual effort. When building a CS system, we must decide how much manual effort is required to solve each of the four CS challenges. This can range from relatively little (for example, combining ratings) to substantial (for example, combining code), and clearly also depends on how much the system is automated. We must decide how to divide the manual effort between the users and the system owners. Some systems ask the users to do relatively little and the owners a great deal. For example, to detect malicious users, the users may simply click a button to report suspicious behaviors, whereas the owners must carefully examine all relevant evidence to determine if a user is indeed malicious. Some systems do the reverse. For example, most of the manual burden of merging Wikipedia edits falls on the users (who are currently editing), not the owners.

Role of human users. We consider four basic roles of humans in a CS system. *Slaves*: humans help solve the problem in a divide-and-conquer fashion, to minimize the resources (for example, time, effort) of the owners. Examples are ESP and finding a missing boat in satellite images using Mechanical Turk. *Perspective providers*: humans contribute different perspectives, which when combined often produce a better solution (than with a single human). Examples are reviewing books and aggregating user bets to make predictions.[29] *Content providers*: humans contribute self-generated content (for example, videos on YouTube, images on Flickr). *Component providers*: humans function as components in the target artifact, such as a social network, or simply just a community of users (so that the owner can, say, sell ads). Humans often play multiple roles within a single CS system (for example, slaves, perspective providers, and content providers in Wikipedia). It is important to know these roles because that may determine how to recruit. For example, to use humans as perspective providers, it is important to recruit a diverse crowd where each human can make independent decisions, to avoid "group think."[29]

Standalone versus piggyback. When building a CS system, we may decide to piggyback on a well-established system, by exploiting traces that users leave in that system to solve our target problem. For example, Google's "Did you mean" and Yahoo's Search Assist utilize the search log and user clicks of a search engine to correct spelling mistakes. Another system may exploit user purchases in an online bookstore (Amazon) to recommend books. Unlike standalone systems, such piggyback systems do not have to solve the challenges of recruiting users and deciding what they can do. But they still have to decide how to evaluate users and their inputs (such as traces in this case), and to combine such inputs to solve the target problem.

A sample of basic CS system types on the World-Wide Web.

Nature of Collaboration	Architecture	Must recruit users?	What users do?	Examples	Target Problems	Comments
Explicit	Standalone	Yes	Evaluating ► review, vote, tag	► reviewing and voting at Amazon, tagging Web pages at del.ici.ous.com and Google Co-op	Evaluating a collection of items (e.g., products, users)	Humans as perspective providers. No or loose combination of inputs.
			Sharing ► items ► textual knowledge ► structured knowledge	► Napster, YouTube, Flickr, CPAN, programmableweb.com ► Mailing lists, Yahoo! Answers, QUIQ, ehow.com, Quora ► Swivel, Many Eyes, Google Fusion Tables, Google Base, bmrb.wisc.edu, galaxyzoo, Piazza, Orchestra	Building a (distributed or central) collection of items that can be shared among users.	Humans as content providers. No or loose combination of inputs.
			Networking	► LinkedIn, MySpace, Facebook	Building social networks	Humans as component providers. Loose combination of inputs.
			Building artifacts ► software ► textual knowledge bases ► structured knowledge bases ► systems ► others	► Linux, Apache, Hadoop ► Wikipedia, openmind, Intellipedia, ecolicommunity ► Wikipedia infoboxes/DBpedia, IWP, Google Fusion Tables, YAGO-NAGA, Cimple/DBLife ► Wikia Search, mahalo, Freebase, Eurekster ► newspaper at Digg.com, Second Life	Building physical artifacts	Humans can play all roles. Typically tight combination of inputs. Some systems ask both humans and machines to contribute.
			Task execution	► Finding extraterrestrials, elections, finding people, content creation (e.g., Demand Media, Associated Content)	Possibly any problem	
Implicit	Standalone	Yes	► play games with a purpose ► bet on prediction markets ► use private accounts ► solve captchas ► buy/sell/auction, play massive multiplayer games	► ESP ► intrade.com, Iowa Electronic Markets ► IMDB private accounts ► recaptcha.net ► eBay, *World of Warcraft*	► labeling images ► predicting events ► rating movies ► digitizing written text ► building a user community (for purposes such as charging fees, advertising)	Humans can play all roles. Input combination can be loose or tight.
	Piggyback on another system	No	► keyword search ► buy products ► browse Web sites	► Google, Microsoft, Yahoo ► recommendation feature of Amazon ► adaptive Web sites (e.g., Yahoo! front page)	► spelling correction, epidemic prediction ► recommending products ► reorganizing a Web site for better access	Humans can play all roles. Input combination can be loose or tight.

SAMPLE CS SYSTEMS ON THE WEB

Building on this discussion of CS dimensions, we now focus on CS systems on the Web, first describing a set of basic system types, and then showing how deployed CS systems often combine multiple such types.

The accompanying table shows a set of basic CS system types. The set is not meant to be exhaustive; it shows only those types that have received most attention. From left to right, it is organized by collaboration, architecture, the need to recruit users, and then by the actions users can take. We now discuss the set, starting with explicit systems.

Explicit Systems: These standalone systems let users collaborate explicitly. In particular, users can evaluate, share, network, build artifacts, and execute tasks. We discuss these systems in turn.

Evaluating: These systems let users evaluate "items" (for example, books, movies, Web pages, other users) using textual comments, numeric scores, or tags.10

Sharing: These systems let users share "items" such as products, services, textual knowledge, and structured knowledge. Systems that share products and services include Napster, YouTube, CPAN, and the site programmableweb. com (for sharing files, videos, software, and mashups, respectively). Systems that share textual knowledge include mailing lists, Twitter, how-to repositories (such as ehow.com, which lets users contribute and search howto articles), Q&A Web sites (such as Yahoo! Answers2), online customer support systems (such as QUIQ,[22] which powered Ask Jeeves' AnswerPoint, a Yahoo! Answers-like site). Systems that share structured knowledge (for example, relational, XML, RDF data) include Swivel, Many Eyes, Google Fusion Tables, Google Base, many escience Web sites (such as bmrb.wisc. edu, galaxyzoo.org), and many peer-to-peer systems developed in the Semantic Web, database, AI, and IR communities (such as Orchestra[8,27]). Swivel, for example, bills itself as the "YouTube of structured data," which lets users share, query, and visualize census- and voting data, among others. In general, sharing systems can be central (such as YouTube, ehow, Google Fusion Tables, Swivel) or distributed, in a peer-to-peer fashion (such as Napster, Orchestra).

Networking: These systems let users collaboratively construct a large social network graph, by adding nodes and edges over time (such as homepages, friendships). Then they exploit the graph to provide services (for example, friend updates, ads, and so on). To a lesser degree, blogging systems are also networking systems in that bloggers often link to other bloggers.

A key distinguishing aspect of systems that evaluate, share, or network is that they do not merge user inputs, or do so automatically in relatively simple fashions. For example, evaluation systems typically do not merge textual user reviews. They often merge user inputs such as movie ratings, but do so automatically using some formulas. Similarly, networking systems automatically merge user inputs by adding them as nodes and edges to a social network graph. As a result, users of such systems do not need (and, in fact, often are not allowed) to edit other users' input.

Building Artifacts: In contrast, systems that let users build artifacts such as Wikipedia often merge user inputs tightly, and require users to edit and merge one another's inputs. A well-known artifact is software (such as Apache, Linux, Hadoop). Another popular artifact is textual knowledge bases (KBs). To build such KBs (such as Wikipedia), users contribute data such as sentences, paragraphs, Web pages, then edit and merge one another's contributions. The knowledge capture (k-cap.org) and AI communities have studied building such KBs for over a decade. A well-known early attempt is openmind,[28] which enlists volunteers to build a KB of commonsense facts (for example, "the sky is blue"). Recently, the success of Wikipedia has inspired many "community wikipedias," such as Intellipedia (for the U.S. intelligence community) and EcoliHub (at ecolicommunity.org, to capture all information about the E. coli bacterium).

Yet another popular target artifact is structured KBs. For example, the set of all Wikipedia infoboxes (that is, attribute-value pairs such as city-name = Madison, state = WI) can be viewed as a structured KB collaboratively created by Wikipedia users. Indeed, this KB has recently been extracted as DBpedia and used in several applications (see dbpedia.org). Freebase.com builds an open structured database, where users can create and populate schemas to describe topics of interest, and build collections of interlinked topics using a flexible graph model of data. As yet another example, Google Fusion Tables (tables.googlelabs.com) lets users upload tabular data and collaborate on it by merging tables from different sources, commenting on data items, and sharing visualizations on the Web.

Several recent academic projects have also studied building structured KBs in a CS fashion. The IWP project35 extracts structured data from the textual pages of Wikipedia, then asks users to verify the extraction accuracy. The Cimple/DBLife project[4,5] lets users correct the extracted structured data, expose it in wiki pages, then add even more textual and structured data. Thus, it builds structured "community wikipedias," whose wiki pages mix textual data with structured data (that comes from an underlying structured KB). Other related works include YAGONAGA, [11] BioPortal,[17] and many recent projects in the Web, Semantic Web, and AI communities.[1,16,36]

In general, building a structured KB often requires selecting a set of data sources, extracting structured data from them, then integrating the data (for example, matching and merging "David Smith" and "D.M. Smith"). Users can help these steps in two ways. First, they can improve the automatic algorithms of the steps (if any), by editing their code, creating more training data,[17] answering their questions[12,13] or providing feedback on their output.[12,35] Second, users can manually participate in the steps. For example, they can manually add or remove data sources, extract or integrate structured data, or add even more structured data, data not available in the current sources but judged relevant.[5] In addition, a CS system may perform inferences over its KB to infer more structured data. To help this step, users can contribute inference rules and domain knowledge.[25] During all such activities, users can naturally cross-edit and merge one another's contributions, just like in those systems that build textual KBs.

Another interesting target problem is building and improving systems running on the Web. The project Wikia Search (search.wikia.com) lets users build an open source search engine, by contributing code, suggesting URLs to crawl, and editing search result pages (for example, promoting or demoting URLs). Wikia Search was recently disbanded, but similar features (such as editing search pages) appear in other search engines (such as Google, mahalo.com). Freebase lets users create custom browsing and search systems (deployed at Freebase), using the community-curated data and a suite of development tools (such as the Metaweb query language and a hosted development environment). Eurekster.com lets users collaboratively build vertical search engines called *swickis*, by customizing a generic search engine (for example, specifying all URLs the system should crawl). Finally, MOBS, an academic project,[12,13] studies how to collaboratively build data integration systems, those that provide a uniform query interface to a set of data sources. MOBS enlists users to create a crucial system component, namely the semantic mappings (for example, "location" = "address") between the data sources.

In general, users can help build and improve a system running on the Web in several ways. First, they can edit the system's code. Second, the system typically contains a set of internal components (such as URLs to crawl, semantic mappings), and users can help improve these without even touching the system's code (such as adding new URLs, correcting mappings). Third, users can edit system inputs and outputs. In the case of a search engine, for instance, users can suggest that if someone queries for "home equity loan for seniors," the system should also suggest querying for "reverse mortgage." Users can also edit search result pages (such as promoting and demoting URLs, as mentioned earlier). Finally, users can monitor the running system and provide feedback.

We note that besides software, KBs, and systems, many other target artifacts have also been considered. Examples include community newspapers built by asking users to contribute and evaluate articles (such as Digg) and massive multi-player games that build virtual artifacts (such as *Second Life*, a 3D virtual world partly built and maintained by users).

Executing Tasks: The last type of explicit systems we consider is the kind that executes tasks. Examples include finding extraterrestrials, mining for gold, searching for missing people,[23,29,30,31] and cooperative debugging (cs.wisc.edu/cbi, early work of this project received the ACM Doctoral Dissertation Award in 2005). The 2008 election is a well-known example, where the Obama team ran a large online CS operation asking numerous volunteers to help mobilize voters. To apply CS to a task, we must find task parts that can be "crowdsourced," such that each user can make a contribution and the contributions in turn can be combined to solve the parts. Finding such parts and combining user contributions are often task specific. Crowdsourcing the parts, however, can be fairly general, and platforms have been developed to assist that process. For example, Amazon's Mechanical Turk can help distribute pieces of a task to a crowd of users (and several recent interesting toolkits have even been developed for using Mechanical Turk[13,37]). It was used recently to search for Jim Gray, a database researcher lost at sea, by asking volunteers to examine pieces of satellite images for any sign of Jim Gray's boat.[18]

Implicit Systems: As discussed earlier, such systems let users collaborate implicitly to solve a problem of the system owners. They fall into two groups: standalone and piggyback.

A standalone system provides a service such that when using it users implicitly collaborate (as a side effect) to solve a problem. Many such systems exist, and the table here lists a few representative examples. The ESP game[32] lets users play a game of guessing common words that describe images (shown independently to each user), then uses those words to label images. Google Image Labeler builds on this game, and many other "games with a purpose" exist.[33] Prediction markets[23,29] let users bet on events (such as elections, sport events), then aggregate the bets to make predictions. The intuition is that the "collective wisdom" is often accurate (under certain conditions)[31] and that this helps incorporate inside information available from users. The Internet Movie Database (IMDB) lets users import movies into private accounts (hosted by IMDB). It designed the accounts such that users are strongly motivated to rate the imported movies, as doing so bring many private benefits (such as they can query to find all imported action movies rated at least 7/10, or the system can recommend action movies highly rated by people with similar taste). IMDB then aggregates all private ratings to obtain a public rating for each movie, for the benefit of the public. reCAPTCHA asks users to solve captchas to prove they are humans (to gain access to a site), then leverages the results for digitizing written text.[34] Finally, it can be argued that the *target problem* of many systems (that provide user services) is simply to *grow a large community of users*, for various reasons (such as personal satisfaction, charging subscription fees, selling ads, selling the systems to other companies). Buy/sell/auction websites (such as eBay) and massive multiplayer games (such as *World of Warcraft*) for instance fit this description. Here, by simply joining the system, users can be viewed as implicitly collaborating to solve the target problem (of growing user communities).

The second kind of implicit system we consider is a piggyback system that exploits the user traces of yet another system (thus, making the users of this latter system implicitly collaborate) to solve a problem. For example, over time many piggyback CS systems have been built on top of major search engines, such as Google, Yahoo!, and Microsoft. These systems exploit the traces of search engine users (such as search logs, user clicks) for a wide range of tasks (such as spelling correction, finding synonyms, flu epidemic prediction, and keyword generation for ads[6]). Other examples include exploiting user purchases to recommend products,[26] and exploiting click logs to improve the presentation of a Web site.[19]

CS SYSTEMS ON THE WEB

We now build on basic system types to discuss deployed CS systems on the Web. Founded on static HTML pages, the Web soon offered many interactive services. Some services serve machines (such as DNS servers, Google Map API server), but most serve humans. Many such services do not need to recruit users (in the sense that the more the better). Examples include pay-parking-ticket services (for city residents) and room-reservation services. (As noted, we call these crowd management systems). Many services, however, face CS challenges, including the need to grow large user bases. For example, online stores such as Amazon want a growing user base for their services, to maximize profits, and startups such as epinions.com grow their user bases for advertising. They started out as primitive CS systems, but quickly improved over time with additional CS features (such as reviewing, rating, networking). Then around 2003, aided by the proliferation of social software (for example, discussion groups, wiki, blog), many full-fledged CS systems (such as Wikipedia, Flickr, YouTube, Facebook, MySpace) appeared, marking the arrival of Web 2.0. This Web is growing rapidly, with many new CS systems being developed and non-CS systems adding CS features.

These CS systems often combine multiple basic CS features. For example, Wikipedia primarily builds a textual KB. But it also builds a structured KB (via infoboxes) and hosts many knowledge sharing forums (for example, discussion groups). YouTube lets users both share and evaluate videos. Community portals often combine all CS features discussed so far. Finally, we note that the Semantic Web, an ambitious attempt to add structure to the

Web, can be viewed as a CS attempt to share structured data, and to integrate such data to build a Web-scale structured KB. The World-Wide Web itself is perhaps the largest CS system of all, encompassing everything we have discussed.

CHALLENGES AND SOLUTIONS

Here, we discuss the key challenges of CS systems:

How to recruit and retain users? Recruiting users is one of the most important CS challenges, for which five major solutions exist. First, we can *require users* to make contributions if we have the authority to do so (for example, a manager may require 100 employees to help build a company-wide system). Second, we can *pay users*. Mechanical Turk for example provides a way to pay users on the Web to help with a task. Third, we can *ask for volunteers*. This solution is free and easy to execute, and hence is most popular. Most current CS systems on the Web (such as Wikipedia, YouTube) use this solution. The downside of volunteering is that it is hard to predict how many users we can recruit for a particular application.

The fourth solution is *to make users pay for service*. The basic idea is to require the users of a system *A* to "pay" for using *A*, by contributing to a CS system *B*. Consider for example a blog website (that is, system *A*), where a user *U* can leave a comment only after solving a puzzle (called a captcha) to prove that *U* is a human. As a part of the puzzle, we can ask *U* to retype a word that an OCR program has failed to recognize (the "payment"), thereby contributing to a CS effort on digitizing written text (that is, system B). This is the key idea behind the reCAPTCHA project.[34] The MOBS project[12,13] employs the same solution. In particular, it ran experiments where a user *U* can access a Web site (such as a class homepage) only after answering a relatively simple question (such as, is string "1960" in "born in 1960" a birth date?). MOBS leverages the answers to help build a data integration system. This solution works best when the "payment" is unintrusive or cognitively simple, to avoid deterring users from using system A.

The fifth solution is to *piggyback on the user traces* of a well-established system (such as building a spelling correction system by exploiting user traces of a search engine, as discussed previously). This gives us a steady stream of users. But we must still solve the difficult challenge of determining how the traces can be exploited for our purpose.

Once we have selected a recruitment strategy, we should consider how to further encourage and retain users. Many *encouragement and retention (E&R) schemes* exist. We briefly discuss the most popular ones. First, we can provide *instant gratification*, by immediately showing a user how his or her contribution makes a difference.[16] Second, we can provide an *enjoyable experience* or a *necessary service*, such as game playing (while making a contribution). [32] Third, we can provide ways to *establish, measure, and show fame/trust/ reputation*.[7,13,24,25] Fourth, we can set up competitions, such as showing top rated users. Finally, we can provide *ownership situations*, where a user may feel he or she "owns" a part of the system, and thus is compelled to "cultivate" that part. For example, zillow.com displays houses and estimates their market prices. It provides a way for a house owner to claim his or her house and provide the correct data (such as number of bedrooms), which in turn helps improve the price estimation.

These E&R schemes apply naturally to volunteering, but can also work well for other recruitment solutions. For example, after *requiring* a set of users to contribute, we can still provide instant gratification, enjoyable experience, fame management, and so on, to maximize user participation. Finally, we note that deployed CS systems often employ a mixture of recruitment methods (such as bootstrapping with "requirement" or "paying," then switching to "volunteering" once the system is sufficiently "mature").

What contributions can users make? In many CS systems the kinds of contributions users can make are somewhat limited. For example, to evaluate, users review, rate, or tag; to share, users add items to a central Web site; to network, users link to other users; to find a missing boat in satellite images, users examine those images.

In more complex CS systems, however, users often can make a far wider range of contributions, from simple low-hanging fruit to cognitively complex ones. For example, when building a structured KB, users can add a URL, flag incorrect data, and supply attribute-value pairs (as low-hanging fruit).[3,5] But they can also supply inference rules, resolve controversial issues, and merge conflicting inputs (as cognitively complex contributions).[25] The challenge is to define this range of possible contributions (and design the system such that it can gather a critical crowd of such contributions).

Toward this goal, we should consider four important factors. First, how *cognitively demanding* are the contributions? A CS system often has a way to classify users into groups, such as guests, regulars, editors, admins, and "dictators." We should take care to design cognitively appropriate contribution types for different user groups.

Low-ranking users (such as guests, regulars) often want to make only "easy" contributions (such as answering a simple question, editing one to two sentences, flagging an incorrect data piece). If the cognitive load is high, they may be reluctant to participate. High-ranking users (such as editors, admins) are more willing to make "hard" contributions (such as resolving controversial issues).

Second, what should be the *impact* of a contribution? We can measure the potential impact by considering how the contribution potentially affects the CS system. For example, editing a sentence in a Wikipedia page largely affects only that page, whereas revising an edit policy may potentially affect million[s] of pages. As another example, when building a structured KB, flagging an incorrect data piece typically has less potential impact than supplying an inference rule, which may be used in many parts of the CS system. Quantifying the potential impact of a contribution type in a complex CS system may be difficult.[12,13] But it is important to do so, because we typically have far fewer high-ranking users such as editors and admins (than regulars, say). To maximize the total contribution of these few users, we should ask them to make potentially high-impact contributions whenever possible.

Third, what about *machine contributions?* If a CS system employs an algorithm for a task, then we want human users to make contributions that are easy for humans, but difficult for machines. For example, examining textual and image descriptions to decide if two products match is relatively easy for humans but very *difficult for machines.* In short, the CS work should be distributed between human users and machines according to what each of them is best at, in a complementary and synergistic fashion.

Finally, the user interface should make it easy for users to contribute. This is highly non-trivial. For example, how can users easily enter domain knowledge such as "no current living person was born before 1850" (which can be used in a KB to detect, say, incorrect birth dates)? A natural language format (such as in openmind. org) is easy for users, but difficult for machines to understand and use, and a formal language format has the reverse problem. As another example, when building a structured KB, contributing attribute-value pairs is relatively easy (as Wikipedia infoboxes and Freebase demonstrate). But contributing more complex structured data pieces can be quite difficult for naive users, as this often requires them to learn the KB schema, among others.[5]

How to combine user contributions? Many CS systems do not combine contributions, or do so in a loose fashion. For example, current evaluation systems do not combine reviews, and combine numeric ratings using relatively simple formulas. Networking systems simply link contributions (homepages and friendships) to form a social network graph. More complex CS systems, however, such as those that build software, KBs, systems, and games, combine contributions more tightly. Exactly how this happens is application dependent. Wikipedia, for example, lets users manually merge edits, while ESP does so automatically, by waiting until two users agree on a common word.

No matter how contributions are combined, a key problem is to decide what to do if users differ, such as when three users assert "A" and two users "not A." Both automatic and manual solutions have been developed for this problem. Current automatic solutions typically combine contributions weighted by some user scores. The work[12,13] for example lets users vote on the correctness of system components (the semantic mappings of a data integration systems in this case[20]), then combines the votes weighted by the trustworthiness of each user. The work[25] lets users contribute structured KB fragments, then combines them into a coherent probabilistic KB by computing the probabilities that each user is correct, then weighting contributed fragments by these probabilities.

Manual dispute management solutions typically let users fight and settle among themselves. Unresolved issues then percolate up the user hierarchy. Systems such as Wikipedia and Linux employ such methods. Automatic solutions are more efficient. But they work only for relatively simple forms of contributions (such as voting), or forms that are complex but amenable to algorithmic manipulation (such as structured KB fragments). Manual solutions are still the currently preferred way to combine "messy" conflicting contributions.

To further complicate the matter, sometimes not just human users, but machines also make contributions. Combining such contributions is difficult. To see why, suppose we employ a machine *M* to help create Wikipedia infoboxes. [35] Suppose on Day 1 *M* asserts population = 5500 in a city infobox. On Day 2, a user *U* may correct this into population = 7500, based on his or her knowledge. On Day 3, however, *M* may have managed to process more Web data, and obtained higher confidence that population = 5500 is indeed correct. Should *M* override *U's* assertion? And if so, how can *M* explain its reasoning to *U*? The main problem here is it is difficult for a machine to enter into a manual dispute with a human user. The currently preferred method is for *M* to alert *U*, and then leave it up to *U* to decide what to do. But this method clearly will not scale with the number of conflicting contributions.

How to evaluate users and contributions? CS systems often must manage malicious users. To do so, we can use a combination of techniques that block, detect, and deter. First, we can block many malicious users by limiting who can make what kinds of contributions. Many e-science CS systems, for example, allow anyone to submit data, but only certain domain scientists to clean and merge this data into the central database.

Second, we can detect malicious users and contributions using a variety of techniques. Manual techniques include monitoring the system by the owners, distributing the monitoring workload among a set of trusted users, and enlisting ordinary users (such as flagging bad contributions on message boards). Automatic methods typically involve some tests. For example, a system can ask users questions for which it already knows the answers, then use the answers of the users to compute their reliability scores.[13,34] Many other schemes to compute users' reliability/trust/fame/reputation have been proposed.[9,26]

Finally, we can deter malicious users with threats of "punishment." A common punishment is banning. A newer, more controversial form of punishment is "public shaming," where a user *U* judged malicious is publicly branded as a malicious or "crazy" user for the rest of the community (possibly without *U's* knowledge). For example, a chat room may allow users to rate other users. If the (hidden) score of a user *U* goes below a threshold, other users will only see a mechanically garbled version of *U's* comments, whereas *U* continues to see his or her comments exactly as written.

No matter how well we manage malicious users, malicious contributions often still seep into the system. If so, the CS system must find a way to undo those. If the system does not combine contributions (such as reviews) or does so only in a loose fashion (such as ratings), undoing is relatively easy. If the system combines contributions tightly, but keeps them localized, then we can still undo with relatively simple logging. For example, user edits in Wikipedia can be combined extensively within a single page, but kept localized to that page (not propagated to other pages). Consequently, we can undo with page-level logging, as Wikipedia does. However, if the contributions are pushed deep into the system, then undoing can be very difficult. For example, suppose an inference rule *R* is contributed to a KB on Day 1. We then use *R* to infer many facts, apply other rules to these facts and other facts in the KB to infer more facts, let users edit the facts extensively, and so on. Then on Day 3, should *R* be found incorrect, it would be very difficult to remove *R* without reverting the KB to its state on Day 1, thereby losing all good contributions made between Day 1 and Day 3.

At the other end of the user spectrum, many CS systems also identify and leverage influential users, using both manual and automatic techniques. For example, productive users in Wikipedia can be recommended by other users, promoted, and given more responsibilities. As another example, certain users of social networks highly influence buy/sell decisions of other users. Consequently, some work has examined how to automatically identify these users, and leverage them in viral marketing within a user community.[24]

CONCLUSION

We have discussed CS systems on the World-Wide Web. Our discussion shows that crowdsourcing can be applied to a wide variety of problems, and that it raises numerous interesting technical and social challenges. Given the success of current CS systems, we expect that this emerging field will grow rapidly. In the near future, we foresee three major directions: more generic platforms, more applications and structure, and more users and complex contributions.

First, the various systems built in the past decade have clearly demonstrated the value of crowdsourcing. The race is now on to move beyond building individual systems, toward building general CS platforms that can be used to develop such systems quickly.

Second, we expect that crowdsourcing will be applied to ever more classes of applications. Many of these applications will be formal and structured in some sense, making it easier to employ automatic techniques and to coordinate them with human users.[37–40] In particular, a large chunk of the Web is about data and services. Consequently, we expect crowdsourcing to build structured databases and structured services (Web services with formalized input and output) will receive increasing attention.

Finally, we expect many techniques will be developed to engage an ever broader range of users in crowdsourcings, and to enable them, especially naïve users, to make increasingly complex contributions, such as creating software programs and building mashups (without writing any code), and specifying complex structured data pieces (without knowing any structured query languages).

References

1. AAAI-08 Workshop. Wikipedia and artificial intelligence: An evolving synergy, 2008.

2. Adamic, L.A., Zhang, J., Bakshy, E. and Ackerman, M.S. Knowledge sharing and Yahoo answers: Everyone knows something. In Proceedings of WWW, 2008.

3. Chai, X., Vuong, B., Doan, A. and Naughton, J.F. Efficiently incorporating user feedback into information extraction and integration programs. In Proceedings of SIGMOD, 2009.

4. The Cimple/DBLife project; http://pages.cs.wisc.edu/~anhai/projects/cimple.

5. DeRose, P., Chai, X., Gao, B.J., Shen, W., Doan, A., Bohannon, P. and Zhu, X. Building community Wikipedias: A machine-human partnership approach. In Proceedings of ICDE, 2008.

6. Fuxman, A., Tsaparas, P., Achan, K. and Agrawal, R. Using the wisdom of the crowds for keyword generation. In Proceedings of WWW, 2008.

7. Golbeck, J. Computing and applying trust in Web-based social network, 2005. Ph.D. Dissertation, University of Maryland.

8. Ives, Z.G., Khandelwal, N., Kapur, A., and Cakir, M. Orchestra: Rapid, collaborative sharing of dynamic data. In Proceedings of CIDR, 2005.

9. Kasneci, G., Ramanath, M., Suchanek, M. and Weiku, G. The yago-naga approach to knowledge discovery. SIGMOD Record 37, 4, (2008), 41-47.

10. Koutrika, G., Bercovitz, B., Kaliszan, F., Liou, H. and Garcia-Molina, H. Courserank: A closed-community social system through the magnifying glass. In The 3rd Int'l AAAI Conference on Weblogs and Social Media (ICWSM), 2009.

11. Little, G., Chilton, L.B., Miller, R.C. and Goldman, M. Turkit: Tools for iterative tasks on mechanical turk, 2009. Technical Report. Available from glittle.org.

12. McCann, R., Doan, A., Varadarajan, V., and Kramnik, A. Building data integration systems: A mass collaboration approach. In WebDB, 2003.

13. McCann, R., Shen, W. and Doan, A. Matching schemas in online communities: A Web 2.0 approach. In Proceedings of ICDE, 2008.

14. McDowell, L., Etzioni, O., Gribble, S.D., Halevy, A.Y., Levy, H.M., Pentney, W., Verma, D. and Vlasseva, S. Mangrove: Enticing ordinary people onto the semantic web via instant gratification. In Proceedings of ISWC, 2003.

15. Mihalcea, R. and Chklovski, T. Building sense tagged corpora with volunteer contributions over the Web. In Proceedings of RANLP, 2003.

16. Noy, N.F., Chugh, A. and Alani, H. The CKC challenge: Exploring tools for collaborative knowledge construction. IEEE Intelligent Systems 23, 1, (2008) 64-68.

17. Noy, N.F., Griffith, N. and Munsen, M.A. Collecting community-based mappings in an ontology repository. In Proceedings of ISWC, 2008.

18. Olson, M. The amateur search. SIGMOD Record 37, 2 (2008), 21-24.

19. Perkowitz, M. and Etzioni, O. Adaptive web sites. Comm. ACM 43, 8 (Aug. 2000).

20. Rahm, E. and Bernstein, P.A. A survey of approaches to automatic schema matching. VLDB J. 10, 4, (2001), 334-350.

21. Ramakrishnan, R. Collaboration and data mining, 2001. Keynote talk, KDD.

22. Ramakrishnan, R., Baptist, A., Ercegovac, A., Hanselman, M., Kabra, N., Marathe, A. and Shaft, U. Mass collaboration: A case study. In Proceedings of IDEAS, 2004.

23. Rheingold, H. Smart Mobs. Perseus Publishing, 2003.

24. Richardson, M. and Domingos, P. Mining knowledgesharing sites for viral marketing. In Proceedings of KDD, 2002.

25. Richardson, M. and Domingos, P. Building large knowledge bases by mass collaboration. In Proceedings of K-CAP, 2003.

26. Sarwar, B.M., Karypis, G., Konstan, J.A. and Riedl, J. Item-based collaborative filtering recommendation algorithms. In Proceedings of WWW, 2001.

27. Steinmetz, R. and Wehrle, K. eds. Peer-to-peer systems and applications. Lecture Notes in Computer Science. 3485; Springer, 2005.

28. Stork, D.G. Using open data collection for intelligent software. IEEE Computer 33, 10, (2000), 104-106.

29. Surowiecki, J. The Wisdom of Crowds. Anchor Books, 2005.

30. Tapscott, D. and Williams, A.D. Wikinomics. Portfolio, 2006.

31. Time. Special Issue Person of the year: You, 2006; http://www.time.com/time/magazinearticle/0,9171,1569514,00.html.

32. von Ahn, L. and Dabbish, L. Labeling images with a computer game. In Proc. of CHI, 2004.

33. von Ahn, L. and Dabbish, L. Designing games with a purpose. Comm. ACM 51, 8 (Aug. 2008), 58-67.

34. von Ahn, L., Maurer, B., McMillen, C., Abraham, D. and Blum, M. Recaptcha: Human-based character recognition via Web security measures. Science 321, 5895, (2008), 1465-1468.

35. Weld, D.S., Wu, F., Adar, E., Amershi, S., Fogarty, J., Hoffmann, R., Patel, K. and Skinner, M. Intelligence in Wikipedia. AAAI, 2008.

36. Workshop on collaborative construction, management and linking of structured knowledge (CK 2009), 2009. http://users.ecs.soton.ac.uk/gc3/iswc-workshop.

37. Franklin, M, Kossman, D., Kraska, T, Ramesh, S, and Xin, R. CrowdDB: Answering queries with crowdsourcing. In Proceedings of SIGMOD 2011.

38. Marcus, A., Wu, E. and Madden, S. Crowdsourcing databases: Query processing with people. In Proceedings of CRDR 2011.

39. Parameswaran, A., Sarma, A., Garcia-Molina, H., Polyzotis, N. and Widom, J. Human-assisted graph search: It's okay to ask questions. In Proceedings of VLDB 2011.

40. Parameswaran, A. and Polyzotis, N. Answering queries using humans, algorithms, and databases. In Proceedings of CIDR 2011.

ABOUT THE AUTHORS

AnHai Doan (anhai@cs.wisc.edu) is an associate professor of computer science at the University of Wisconsin-Madison and Chief Scientist at Kosmix Corp.

Raghu Ramakrishnan (ramakris@yahoo-inc.com) is Chief Scientist for Search & Cloud Computing, and a Fellow at Yahoo! Research, Silicon Valley, CA, where he heads the Community Systems group.

Alon Y. Halevy (halevy@google.com) heads the Structured Data Group at Google Research, Mountain View, CA.

From: Anhai Doan, Raghu Ramarkrishnan, & Alon Y. Halevy, "Crowdsourcing: What it Means for Innovation," *Communications of the ACM* 54, no. 4 (2011): 86-96. Used with permission.

13

CHAPTER 13

LEADING PUBLIC & VOLUNTEER ORGANIZATIONS

In this chapter we will focus on the motivation and recognition of volunteers, since volunteer service is essential to CAP in accomplishing our assigned missions. We begin with a quick look at leadership of volunteers, and then move on to management philosophies for organizations. We will do this by taking a look at the structure and behaviors of organizations, with a focus on understanding leadership and management issues that affect public and volunteer organizations.

CAP differs from public and private organizations in that our missions are accomplished by unpaid volunteers. Understanding the motivation of volunteers and harnessing their talents is not the same as managing and directing employees. The author of the first article, "Leadership for Volunteers: The Way It Is and The Way It Could Be," identifies assumptions about working with volunteers that can disrupt effective employment of volunteers. The author also explains how recognition of these assumptions can lead to improved relations with and service to volunteers.

Continuing on this theme, the second reading can be used as a practical checklist for three essential tasks in managing volunteers: recruitment, retention, and recognition. The excerpts from the "Volunteer Management Guidebook" illustrate a number of important concepts of volunteerism, including why people volunteer, how to write volunteer position descriptions, how to close out a project, and rules for recognition.

In the same way that individual volunteers are held to the Core Value of Integrity, organizations can also be held to ethical standards. Unfortunately, we have seen many examples in the news recently of companies that failed to exhibit ethical behavior. Chapter 6 introduced you to the concept of Maslow's hierarchy of personal needs. The authors of the next article, "The Hierarchy of Ethical Values in Nonprofit Organizations: A Framework for an Ethical, Self-Actualized Organization Culture," argue that organizations can promote ethical behavior by ascending and satisfying five levels of an ethical values hierarchy.
One way that organizations can demonstrate ethical behavior is through transparency: open communication and accountability to stakeholders. In the fourth article, "The New Look of Transparency," the author highlights the need for organizations to be open and straightforward in their interactions, especially with and among employees. Though the author's focus is on companies, you can replace the word 'employees' with 'volunteers' to see how the message applies to CAP. Examples of CAP's transparency include an Annual Report to Congress and yearly financial reporting to the public on an IRS Form 990.

In the final article, "Public and Private Management," the author explores the similarities of and differences between management in public organizations and private business. Though the article might seem dated, this report is a classic text that is used by business and management schools around the country even today. As you read about the differences in leading government agencies and private companies, keep in mind that each organization type has both benefits and challenges.

But which type describes CAP? CAP is a non-profit corporation with roughly 150 employees for program management and membership support, 68 unpaid corporate officials, and 61,000+ volunteers. Our funds come from private donations, member dues, and Congressional funds funneled through the Air Force. Our oversight comes from 11 appointed members of a Board of Governors, a national commander and vice commander elected by corporate officers, and Air Force officers and enlisted airmen assigned to CAP-USAF billets. Our organizational structure and missions are determined by Congress and the Air Force. This description indicates that CAP as an organization doesn't clearly fall into either the public or private category, but pulls some characteristics from each.

CHAPTER OUTLINE

This chapter's readings are:

Leadership for Volunteers
Richard Cummins, "Leadership for Volunteers: The Way It Is and The Way It Could Be," *Journal of Extension* 36, no. 5 (1998).

Take Root:
Volunteer Management Guidebook
Corporation for National & Community Service & Hands On Network, "Take Root: Volunteer Management Guidebook," (2010).

The Hierarchy of Ethical Values in Nonprofit Organizations
Ruth Ann Strickland & Shannon K. Vaughan, "The Hierarchy of Ethical Values in Nonprofit Organizations," *Public Integrity* 10, no. 3 (2008): 233-251.

The New Look of Transparency
Kristin Clarke, "Clear: The New Look of Transparency," *Associations Now* (November 2010).

Public & Private Management
Graham T. Allison Jr., "Public & Private Management: ...Alike in All Unimportant Respects?," *Setting Public Management Research Agendas*, OPM Document 127-53-1 (February 1980): 27-38.

CHAPTER GOALS

1. Give examples of effective strategies for leading volunteers.
2. Defend the idea that organizations should be held to high ethical standards.
3. Comprehend key differences in leading public and private organizations.

13.1 Leadership for Volunteers: The Way It Is and The Way It Could Be

By Richard Cummins

OBJECTIVES:

1. Define the term "intrinsic".
2. Define the term "extrinsic".
3. Identify four assumptions that can hinder the success of volunteers in an organization.
4. List four arts that can contribute to the success of volunteer organizations.

ABSTRACT

The failure of volunteer organizations is commonly attributed to a lack of leadership for the organization. The failure problem may be more closely related to unrealistic assumptions rather than the lack of leadership. Identifying common assumptions about organizational goals, volunteer roles, information flow, and feedback is crucial. Addressing those assumptions by learning the arts of active listening, mentoring, public dialogue, and evaluation and reflection is critical to the success of an organization.

For some time, questions have been asked about why some volunteer organizations are more successful than others. By and large, the problem is not with the leadership of the organization. Many talented volunteers bring substantial leadership experience from either the private or the public sector. The problem may be more closely related to unrealistic assumptions regarding the implementation of leadership for organizations.

Through personal experience, four common assumptions regarding leadership for volunteers have emerged. These benchmarks were more a result of armchair observations and hard knocks than the result of research. Research indicates these heuristics, however commonly accepted when working with volunteers, may cause more problems than they cure.

Volunteers are attracted to organizations for a variety of reasons. Generally, the motivations for aligning with others in a voluntary effort can be classified either as intrinsic, that is, doing something for the sake of the activity, or extrinsic, or doing something for an expected payoff. Whichever the case, the volunteer expects to do something. The following generally accepted assumptions may be a source of problems for volunteers willing to work.

Assumption One: Everyone knows what the organization stands for and represents. Volunteers select organizations because of the vision and mission of the organization. In order to fulfill an organization's mission, goals must be clearly articulated to the volunteers. Clearly, volunteers want to do something to help reach the goals and vision of the organization. With the increasing mobility of volunteers, the makeup of an organization will change rapidly and the assumption that everyone knows the mission of the organization is risky. The only way to assure common goals is to frequently share those goals.

Assumption Two: Everyone knows their role. In the work world, employees are usually provided a listing of expectations for their job, such as work standards, appropriate time schedules, authorization capabilities, oversight responsibilities, and reporting protocol. Volunteers have different motivations for voluntary work than paid employees; however, specific guidelines are required in order to have a smooth functioning organization. Role clarification cannot be over-emphasized in volunteer organizations.

Assumption Three: Everyone knows where to get needed information. Volunteers need to know and understand how different parts of a project fit together. Newsletters may give general comments and updates about a project but are usually inadequate regarding specifics about project progress. In addition to the informal lines of communication that develop, a specific reporting mechanism should be established and implemented. Many problems can be avoided when the information flow is unimpeded.

Assumption Four: Everyone gets feedback. It has been said that in Vietnam, the U.S. military did not fight a nine-year war; but rather because of frequent troop changes with no feedback or institutional memory, the U.S. military fought the first year of a war nine times in succession. Volunteers cycle through organizations in much the same way and new recruits are often unaware of previous efforts. Providing feedback to volunteers is critical at all levels of the organization. Special attention is required in order to share previous experiences with current members.

Becoming aware of assumptions and the effects those assumptions have is important in any endeavor. In order to address organizational assumptions, leaders for volunteers should be aware of four arts for sustained involvement. Learning and practicing these arts can contribute to success for volunteers and their chosen organizations.

Art One: Active Listening. Encourage others to talk and search for meaning. Be aware of values of volunteers and strive to meld organizational values and individuals' values. Encourage volunteers to talk about the organization and what they expect from the volunteering experience.

Art Two: Mentoring. Supportively guide others in learning and sharing not only how, but why specific roles are important. Strive to match available skills with volunteers' and organizational needs. Help others solve problems that are holding the organization back.

Art Three: Public Dialogue. Encourage public talk on matters that concern us all. Facilitate interaction to help volunteers gain understanding and appreciation for all segments of a project. Emphasize the free-flow of information.

Art Four: Evaluation and Reflection. Assess and incorporate the lessons we learn through action. Public decision making encourages those expected to implement plans to have ownership of those plans. Encourage new volunteers to make suggestions and avoid suggesting a lockstep method for the organization.

Providing leadership for volunteers can be exhilarating, frustrating, exciting, tedious, rewarding and demanding, all at the same time. Learning how to assess what is and assessing what could be is an important function of leadership for volunteers. Investing time to learn and practice the four arts for sustained involvement can yield substantial results.

ABOUT THE AUTHOR

Richard Cummins
Visiting Assistant Professor
Bush School of Government and Public Service and
Department of Agricultural Education
Texas A&M University
College Station, Texas
r-cummins@tamu.edu

From: Richard Cummins, "Leadership for Volunteers: The Way It Is and The Way It Could Be," *Journal of Extension* 36, no. 5 (1998). Used with permission.

13.2 Take Root: Volunteer Management Guidebook

From the Corporation for National & Community Service and the Hands On Network

OBJECTIVES:

5. Identify three needs that motivate people to volunteer.
6. List the key components of a volunteer position description.
7. Name three goals of volunteer retention.
8. List some rules for volunteer recognition.
9. Name some ways to recognize volunteers in each event category: everyday, intermediate, and large-scale.

CHAPTER 1: VOLUNTEER RECRUITMENT

Overview

Without volunteers, most not-for-profits would cease to exist or would otherwise suffer a drastic reduction in capacity to serve communities and achieve the mission of the organization. Recruitment is the first step in securing volunteer participation in your national service program. This chapter will help you explore the basics of recruitment and how to create a recruitment strategy.

Goals

- Understand volunteer motivation
- Explore basics of recruitment
- Develop a recruitment strategy

Volunteer Motivation

Before you can begin recruiting volunteers for your project, you must first understand who volunteers and why. In a report released in December 2004, the Bureau of Labor Statistics noted Americans' strong commitment to volunteering. Between September 2003 and September 2004, about 64.5 million Americans engaged in volunteer work! Many factors [influence] people to volunteer. Some reasons include:

- They were personally asked.
- An organization with which they are affiliated is participating.
- They have a personal connection to the mission of the project or organization.
- They enjoy the type of work being performed.
- They want to learn new skills.
- They want to meet people.

One study from Independent Sector (2001) reports that 71% of people asked to volunteer, did.

Volunteering is also a great way to develop personal and professional skills. These skills include cultural awareness, creativity, problem solving, and teamwork. Volunteering can also meet motivational needs, as outlined by McClelland and Atkinson's Motivational Theory. According to this theory, people have three separate motivational needs, with one being [predominant]:

Affiliation

The affiliation-motivated person needs personal interaction, works to make friends, likes to get involved with group projects, and needs to be perceived as a "good" person.

Achievement

The achievement-motivated person needs specific goals to work toward, seeks responsibility, sticks to tasks until completed, and sees problems as challenges.

Power

The power-motivated person needs to impact and influence others, can work alone or in a group, can respond to needs of people or programs, and keeps an eye on overall goals of the agency.

Understanding why people volunteer and their motivational needs will help you target your recruitment strategies to engage the volunteers you need to achieve your project goals. While some volunteers may only relate to their own personal reasons for volunteering, you must articulate the relationship between the work of the project and the benefit to either the community or the volunteer. You can convey this and other motivating messages in your recruitment efforts.

Recruitment Basics

Recruitment is the process of enlisting volunteers into the work of the program. Because volunteers give their time only if they are motivated to do so, recruitment is not a process of persuading people to do something they don't want to do. Rather, recruitment should be seen as the process of showing people they can do something they already want to do. People already know that there are problems in the world, that people, the environment and animals need the support of people who care. As a volunteer recruiter it is your job to enroll people into knowing they are that person who cares, give them incentives to volunteer, and point out exactly how they are capable of helping.

Your Program's History, Culture, and Cause
Before you begin to recruit, be sure you understand your national service program's history, culture, and cause. You should be able to answer the following questions:

- How do we typically use our volunteers (committed or not, mostly service days, randomly or regularly)?
- Which programs are successful? Which are/were not?
- With whom have we collaborated? Which of those unions were successful?
- Which were not?
- What publicity, good or bad, has our program received that may affect our recruitment efforts?
- Can we speak knowledgeably about our program's mission/cause?
- Do we feel comfortable speaking to how the projects of the national service program will help achieve the mission?
- Can we clearly articulate to volunteers how their work will contribute to the program's mission and goals?
- Is our workplace open and friendly to volunteers?
- Would we recommend volunteering in our program to close friends and family? Why or why not?

Determining Volunteer Needs

Effective volunteer recruitment begins with a volunteer program that is well planned and executed and that offers meaningful work. Program staff should clarify the work that needs to be done to achieve the goals of the project/program and then segment that work into components that reflect the reality of today's work force.

You will need to consider the type(s) of volunteers you need for your project or program. Think beyond your traditional volunteer base. Do you need someone with many hours to devote to the project, or people who want to serve only one afternoon? Is the project appropriate for children, seniors, or other people with different abilities and needs? Some trends and groups to consider include:

Long-term volunteering
Long-term service provides volunteers the opportunity to commit to a project or program that spans an extended period of time.

Short-term/episodic volunteering
Episodic volunteer opportunities include those that are of short duration and those that occur at regular intervals, such as annual events.

Family volunteering
Family volunteering provides volunteers the opportunity to participate in meaningful service while spending time with their families.

Student volunteering
Through volunteering with schools and youth groups, young people gain valuable knowledge and skills.

Internships
Through internships, fellowships, and apprenticeships, students gain valuable experience while serving the community service organization.

Virtual volunteering
Virtual volunteering allows anyone to contribute time and expertise without ever leaving his or her home.

For many volunteer opportunities, you can work with an advisory team or conduct a survey to identify volunteer assignments that will help advance the goals of the program.

If you are working on a specific service project, you can determine your volunteer needs through developing a task list. Consider what you want to accomplish and the tasks needed; then create a comprehensive list of the assignments and the number of volunteers needed for each task.

Volunteer Position Descriptions

The volunteer position description is a helpful tool. It outlines responsibilities, support, and benefits of specific volunteer opportunities. It also strengthens your recruitment efforts because it defines the assignment, skills, abilities, and interests necessary to perform the task successfully.

A volunteer position description should include the following components:

> ***Title***
> Provide a descriptive title that gives the volunteer a sense of identity. This will also help program staff and other volunteers understand the assigned role.
>
> ***Purpose/objective***
> Use no more than two sentences to describe the specific purpose of the position. If possible, state the purpose in relation to the nonprofit's mission and goals.
>
> ***Location***
> Describe where the person will be working.
>
> **Key responsibilities**
> List the position's major responsibilities. Clearly define what the volunteer is expected to do as part of this assignment.

Qualifications
Clearly list education, experience, knowledge, skills, and age requirements. Also note if the opportunity is accessible to people with disabilities. If a background check is required, it should be indicated here.

Time commitment
Note the length of the assignment, hours per week, and/or other special requirements.

Training/support provided
Define nature and length of all general and position-specific training required for the assignment. Also list resources and other support available to the volunteer.

Benefits
Describe benefits available to volunteer, such a lunch, T-shirt, development opportunities.

Volunteer supervisor and contact information
List the staff person or volunteer leader who will be working most directly with the volunteer and his/her contact information.

Recruitment Strategies

You have determined your volunteer needs and created a task list and/or position descriptions for the assignments. The next step is to create a recruitment strategy to determine whom you will ask to volunteer and how you will ask them.

First, examine the volunteer positions to be filled. Ask yourself these questions:

- Who will be qualified for and interested in this position?
- Who will be able to meet the time commitments?
- Where will you find these people?
- What motivates them to serve?
- What is the best way to approach them?

Now that you have decided on who, you need to start thinking about how to target them.

Remember that different messages will appeal to different audiences, so you will want to use a variety of recruitment methods. You can use targeted recruitment that is focused and addressed to a specific audience where people will have the skills, interests, and availability needed to fill your positions. Broad-based recruitment can be effective for positions requiring minimal training or for when you need a lot of people for a short-term event.

Here are just a few ways of recruiting volunteers:

- The personal ask is always the most compelling!
- Post your volunteer opportunity on the Web, using your program's Web site or another site such as volunteermatch.com.
- Strategically post flyers or brochures in the community.
- Partner with volunteers from a school, corporation, community center, faith-based group, or other non-profit.
- Utilize local media (e.g., newspapers and radio) to spread the word about your volunteer opportunities.
- Network with community groups and leaders.
- Use on-line forums and/or blogs to spread the word.

No matter the volunteer opportunity, you should have some idea of where to look for volunteers in your community. Consider a wide range of individuals and groups that are potential volunteers for your program or project, as well as locations to post flyers and brochures:

- Faith-based groups and/or houses of worship
- Military bases or retired military groups
- Unions and trade workers associations
- Sororities and fraternities
- Teacher's associations
- Retired firefighter, police, and executive associations
- Moms' groups
- Realtors (welcome wagon packages)
- Independent living homes
- Disability services groups
- Scouts, 4-H, Boys & Girls Clubs, or other youth organizations
- Other national service programs
- Grocery store billboards
- Bingo halls
- Doctors' offices
- Public transit stations
- Shopping malls
- Corporate buildings
- Job counseling offices
- Schools
- Salons
- Restaurants
- Newsletters

Don't forget to get permission to display information in specific locations. You may want to ask the owners/managers to attend an orientation so they can better inform interested volunteers who pick up a flyer.

Here are some other tips to build volunteer initiatives:

- Make sure all staff know about the opportunities available for volunteering with your program and where to refer interested volunteers.
- Integrate volunteer management skills into staff training.
- Visit off-site volunteer projects so that the volunteers associate your program with the project.

- Use surveys to find out the interests of volunteers.
- Use colorful descriptions for volunteer positions that are clear and straightforward.
- Try not use the word "volunteer" on marketing pieces. You run the risk of attracting only those who already volunteer or other volunteer managers.

Remember that anyone can be a volunteer. People vary by age, race, ethnicity, religion, sexual orientation, ability, and income. Not all volunteers are the same and not all types of volunteering will appeal to all groups, so have diverse volunteering opportunities available and target recruitment in ways that will appeal to different groups.

Your recruitment strategy is the key to engaging the right number of people with the right skills, interest, and availability for the job. Make sure to plan for a wide variety of volunteers at your project. If you have too few participants, the project will likely go unfinished. If you have too many, some volunteers will have little to do and might feel that their time was ill spent.

The above information is a compilation of materials from Make A Difference, a member organization of Hands On Network; *Volunteer Management* by Steve McCurley and Rick Lynch (1997); http://www.serviceleader.org; and http://www.independentsector.org.

CHAPTER 3: VOLUNTEER RETENTION

{ Chapter 2 is not included in this excerpt }

Overview

The best way to increase your volunteer base is to retain current volunteers. Retention is simply a matter of making volunteers feel good about themselves and their service. It includes motivating volunteers before the project, engaging them during the project, and leading them to reflect on the project. This chapter will provide you with tips for retaining volunteers for your program and projects.

Goals

- Motivate volunteers before the project
- Engage volunteers during the project
- Reflect after the project

Before the Project: Motivation

From the very beginning of volunteers' involvement in your service activities, you should maintain good communication with them. Motivate them to stay interested and involved in your project with a few simple steps:

- Be prompt in your response to phone calls/e-mails. Return volunteer calls or e-mails within 24 hours.
- Be thorough in your explanation of the volunteer duties. Volunteers will be more likely to sign up if they know exactly what they will be doing, and they will know what to expect at the project.
- Use this opportunity to teach potential volunteers about the issue area, the community service organization they will be serving, and the potential impact of the project.
- Use their names often; this helps develop a personal connection.
- Keep the commitments you make. People will not support you if you don't provide information requested, address issues they bring up, and/or miss scheduled appointments.

Continue to be in contact with your team. Keeping volunteers motivated and excited about your project is the best guarantee for success! The more contact you provide, the more engaged your volunteers will be, and the more motivated they will be when they arrive. Also, respond to people's inquiries in a timely and thorough manner.

Make sure to confirm project details with them. Contact volunteers with a phone call or e-mail that:

- Introduces you (or another staff person, partner, or volunteer) as the project leader
- Thanks them for volunteering
- Provides the date and time of the project, service site address, directions, and parking information
- Describes what will occur at the project
- Lets volunteers know what to wear or not wear to the project
- Encourages volunteers to bring supplies they may have
- Tells volunteers whom to contact if they have a change in plans

By communicating all details and project background to volunteers and staying in touch with them frequently, they will begin to create an attachment to the affiliate and the project even before they arrive. Thus they are more likely to show up on the day of the project and want to stay involved with your program for future volunteer opportunities.

During the Project: Engagement

Volunteer management incorporates elements of project and volunteer management. Having a well-planned and well-run project will make the volunteer experience more enjoyable and meaningful, thus they will be more likely to engage in future service. On the day of the project, designate an area for volunteers to "check in." This will allow you to better manage volunteers that attend the project and effectively track their volunteer hours. Welcome volunteers as they arrive. Use nametags at the project and get to know your volunteers. Introduce volunteers to one another to encourage interaction.

Orientation & Training

Have the Project Leader and/or agency representative give a brief (5-10 minute) orientation. The orientation should include an overview of the agency's mission and services, and how volunteer support is contributing to that mission.

Discuss the community issue that is being addressed by the agency and/or the project. This segment could include a brief history of the issue, current statistics, current events related to the issue area (e.g., legislation activity), and other civic engagement opportunities linked to this issue (e.g., advocacy training, future service projects).

The orientation and education portion of the project has many positive impacts on volunteers. It allows them to:

- See the impact they are having on the agency and its clients
- Feel a greater part of a whole, when they see all the services the agency provides
- Better understand the critical needs of the community
- Better understand how to effect change within the issue being addressed

Orientation makes volunteers feel connected to the agency, clients, or their community, and makes their work more meaningful, and in turn makes them more likely to engage in future service.

After the orientation, give a brief outline of the project and what volunteers will be doing during the project, so that everyone knows what to expect and what is expected of them. Also be sure to allow time for training volunteers for any specialized tasks or skills they will need to successfully complete the project.

Utilization

Make sure everyone has something to do. Underutilization is one of the biggest threats to retention. If people do not feel needed, they will not come back.

Balancing Different Personalities

Working with groups can be challenging. Understanding volunteers' personalities can help you position them in different teams of your project so they have the best change of personal success and compatibility with you and other volunteers. Some volunteers want to lead, some want to socialize, some pay attention to details, and others are compassionate and dependable. You may also encounter volunteers who are headstrong, who aren't actively involved, or who complain excessively.

When you are dealing with groups, you are almost guaranteed to encounter clashing personalities. Just remember: opposite personalities can complement one another if they try to understand the other's perspective. Treat every individual with dignity and respect:

- Talk openly and professionally with your volunteer to try to eliminate the problem.
- Consult with another staff person or volunteer leader who can troubleshoot with you on ways to resolve the problem.
- Document any incidents immediately and contact the office if you do not feel you can resolve the problem.
- If a client is causing problem, consult with the agency contact immediately.
- The agency is responsible for managing the clients; you are responsible for managing the volunteers.

It is important to recognize and deal with problem volunteers. You cannot just ignore the problem and expect it to go away. It will affect other volunteers and their experience, and may influence them negatively.

Project Closure

- Always leave time for clean up. Utilize your volunteers to assist with this!
- Review the accomplishments of the day (e.g., number of meals served, walls painted, boxes sorted, etc.) so volunteers have an idea of the impact of their service. Discuss how these accomplishments may have affected the service recipients.
- Reflect on the project (see below).
- Invite volunteers to participate in future service and take future action related to the issue or national service program.

After the Project: Reflection

Reflection is an important part of offering closure to a project. Reflection allows volunteers to stop for a moment, think about what they've accomplished, share their experiences, and offer feedback for future projects or ideas for how they will continue to address the social issue. Reflection is designed to encourage volunteers to examine the project so that they see the impact of their service. Understanding how their service impacts the community will encourage volunteers to be involved in future projects. Reflection can be conducted in many ways. Volunteers can have a group discussion, write about their experience, create a photo-journal of the project, or respond to quotes about service. Be creative and allow for interaction. Encourage each volunteer to contribute to the discussion and make sure that all volunteers have an opportunity to share their thoughts.

Sample questions for reflection discussion:

- What did you learn today—about the agency, your fellow volunteers, the service recipients, or yourself?
- How do you feel about the project? Was it worthwhile? Was it time well spent?
- What would you change about this project?
- Do you plan to take future action related to this issue?
- What other ideas or opinions can you offer this program or project?

Be creative in how you offer reflection activities. Here are a few ideas:

Web of Service: Have the group stand in a circle. Holding the end of a ball of string, hand the ball off to another participant. Ask him or her to reflect on a particular question (e.g., what was something new you learned today?). Once she has answered the question, she should hold on to her piece of the string and pass the ball onto someone else. Continue the process until everyone has reflected on the question and has a section of string in his or her hands. When completed, you should have something that looks like a web. When everyone has answered, make some points about the interconnectedness of people, how they are all part of the solution, for if one person had not contributed to their service projects the outcome would've been different. End with another question and have the volunteers retrace their steps passing the string in reverse order.

Talking Object: Gather volunteers in a circle or semi-circle, offer a "talking object" to pass around the circle as people answer reflection questions.

One to Three Words: Each person shares one to three words to describe the service activity or how he/she feels about the service activity or anything else regarding the project.

Poetry: Have volunteers write a sentence about the project. Gather these sentences. Then after a few months/weeks, compile the sentence into a poem or story.

Quotes of Service: Quotes are useful to encourage volunteer reflection. Project leaders can read the quote and ask for a response or simply let the volunteers silently reflect on the words as they part from each other.

"Without community service, we would not have a strong quality of life. It's important to the person who serves as well as the recipient. It's the way in which we ourselves grow and develop..."
— Dr. Dorothy I. Height president and CEO,
National Council of Negro Women

"No joy can equal the joy serving others." —Sai Baba

"When you give to others, you speak a silent but audible thank you. Appreciation for others and ourselves is life and spirit for each and every one of us."
— Kara "Cherry" Whitaker, 14 years old, Ohio

"You make a living by what you get. You make a life by what you give."
—Winston Churchill

Be sure to thank volunteers for their efforts and inform them of upcoming projects. For volunteers who frequently return to your projects or who show leadership or desire additional responsibilities, delegate tasks to get them more involved and connected to your project or to allow them to enhance skills. Let them make phone calls, send e-mails, be in charge of specific tasks, etc. Encourage them to become a Volunteer Leader. Retaining volunteers is essential to the success of your program. From project beginning to end, volunteers need to feel good about themselves and their service. You can accomplish this through motivation, engagement, and reflection. Many times retention and recognition are so closely related that they are actually interdependent on one another. The next chapter will offer many tips and suggestions for recognizing volunteers for their service efforts.

The above information is a compilation of materials from Make A Difference, a 501(c)(3); *Volunteer Management* by Steve McCurley and Rick Lynch (1997); http://www.e-volunteerism. com; and http://www.hiresuccess.com/pplus-3.htm.

CHAPTER 4: VOLUNTEER RECOGNITION

Overview

Recognition is a key component of volunteer management. Volunteers need to know that their service has made an impact and that they are appreciated by the community, fellow volunteers, and program staff. This chapter will offer ideas for creative ways to tailor recognition to meet your individual volunteers' needs.

Goals

- Recognize volunteers

- Understand rules for recognition
- Explore tips and tools for recognition

Recognizing Volunteers

Recognition makes volunteers feel appreciated and valued. If volunteers don't feel like their contribution is valuable or necessary, they won't return. Volunteer recognition can take many forms, from a simple thank-you card to a large annual event. An ideal recognition system makes use of many different procedures, to have something for every volunteer and to keep it personal and meaningful.

Matching Recognition to Types of Volunteers

Volunteers have different personalities, are motivated to serve for different reasons, and serve in different ways. Therefore, you should use a variety of recognition methods for your volunteers. Tailor your recognition to individual volunteers, to make it most effective and meaningful.

By Motivational Orientation

Think about recognition that is appropriate for volunteers with different motivational needs.

Achievement-oriented volunteers:

- Ideal result of recognition is additional training or more challenging tasks.
- Subject for recognition is best linked to a very specific accomplishment.
- Phrasing of recognition should include "Best" or "Most" awards.
- Recognition decision should include checkpoints or records.
- Awardees should be selected by co-workers.

Affiliation-oriented volunteers:

- Recognition should be given at a group event.
- Recognition should be given in presence of peers, family, or other bonded groups.
- Recognition should have a personal touch.
- Recognition should be organizational in nature, given by organization.
- Recognition should be voted by peers.

Power-oriented volunteers:

- Key aspect of recognition is "promotion," conveying greater access to authority or information.
- Recognition should be commendation from "Names."
- Recognition should be announced to community at large, put in newspaper, etc.
- Recognition decision should be made by the organization's leadership.

By Style of Volunteering

You should also vary recognition for volunteers who serve one afternoon [vs. those who serve one year.]

Long-term volunteer:

- Recognition with and by the group
- Recognition items make use of group symbols
- Recognition entails greater power, involvement, information about the organization
- Presenter of recognition is a person in authority

Short-term (episodic) volunteer:

- Recognition is given in immediate work unit or social group
- Recognition is "portable" – something the volunteers can take with them when they leave, such as a present, photograph, or other memorabilia
- Presenter is either the immediate supervisor or the client

Informal vs. Formal Recognition

Day-to-day recognition is the most effective because it is much more frequent than a once-a-year banquet and helps to establish good working relationships.

Formal recognition includes awards, certificates, plaques, pins, and recognition dinners or receptions to honor volunteer achievement. They are helpful mainly in satisfying the needs of the volunteer who has a need for community approval, but have little impact (or occasionally a negative impact) on volunteers whose primary focus is helping the clientele. These volunteers may feel more motivated and honored by a system which recognizes the achievements of their clients and the contribution the volunteer has made towards this achievement.

When determining whether to establish a formal recognition, consider the following:

- Is this being done to honor the volunteer, or so the staff can feel involved and can feel that they have shown their appreciation for volunteers?
- Is it real and not stale or mechanical?
- Does it fit? Would the volunteers feel better if you spent the money on the needs of the clients rather than on an obligatory luncheon?
- Can you make it a sense of celebration and builder of team identity?

Goals for a Recognition Event

Educate

- Educate everyone attending about the scope, meaning, and value of volunteer services to your organization.
- Report the outcomes of volunteer effort.

- Gain publicity for the organization and the volunteer program.
- Gain respect for the volunteer program and the director of it.

Inspire

- Recommit (re-enthuse) volunteers for the coming year of work recruit.
- [Find] new volunteers for vacant positions.
- Challenge all volunteers through recognition of the accomplishments of a selected few.

Recognize

- Say thank you for everything and anything volunteered to the organization during the past year, and make sure everyone volunteering during the past year feels appreciated.
- Acknowledge the contributions of some paid staff supervisors to the success of volunteers.

Have fun!

- Allow volunteers, and staff, a chance to have fun and meet each other.

Creative Recognition

Volunteer recognition does not have to cost a lot, and there are many alternatives to the traditional annual recognition banquet. Use your imagination and think outside the box to come up with some fun, inexpensive ideas that are fun for volunteers.

Rules for Recognition

Recognize . . . or else — The need for recognition is very important to most people. If volunteers do not get recognition for productive participation, it is likely that they will feel unappreciated and may stop volunteering with your program.

Give it frequently — Recognition has a short shelf life. Its effects start to wear off after a few days, and after several weeks of not hearing anything positive, volunteers start to wonder if they are appreciated. Giving recognition once a year at a recognition banquet is not enough.

Give it via a variety of methods — One of the implications of the previous rule is that you need a variety of methods of showing appreciation to volunteers.

Give it honestly — Don't give praise unless you mean it. If you praise substandard performance, the praise you give to others for good work will not be valued. If a volunteer is performing poorly, you might be able to give him honest recognition for his effort or for some personality trait.

Give it to the person, not the work — This is a subtle but important distinction. If volunteers organize a fund-raising event, for example, and you praise the event without mentioning who organized it, the volunteers may feel some resentment. Make sure you connect the volunteer's name to it. It is better to say "John, Betty, and Megan did a great job of organizing this event" than to say "This event was very well-organized."

Give it appropriately to the achievement — Small accomplishments should be praised with low-effort methods, large accomplishments should get something more. For example, if a volunteer tutor teaches a child to spell "cat" today we could say "Well done!" If she writes a grant that doubles our funding, a banner lauding her accomplishment might be more appropriate.

Give it consistently — If two volunteers are responsible for similar achievements, they ought to get similar recognition. If one gets her picture in the lobby and another gets an approving nod, the latter may feel resentment. This does not mean that the recognition has to be exactly the same but that it should be the result of similar effort on your part.

Give it on a timely basis — Praise for work should come as soon as possible after the achievement. Don't save up your recognition for the annual banquet. If a volunteer has to wait months before hearing any word of praise, she may develop resentment for lack of praise in the meantime.

Give it in an individualized fashion — Different people like different things. One might respond favorably to football tickets, while another might find them useless. Some like public recognition; others find it embarrassing. In order to provide effective recognition, you need to get to know your volunteers and what they will respond to positively.

Give it for what you want more of — Too often your staff pays most attention to volunteers who are having difficulty. Unfortunately, this may result in ignoring good performers. We are not suggesting that you ignore sub-par volunteers, just that you make sure that you praise the efforts of those who are doing a good job.

Tips and Tools for Recognition

Here are some easy, everyday ways to recognize volunteers:

- Use e-mail to send thank you letters/messages.
- Send postcards or thank you cards to volunteers after they attend a project.

- Send a birthday card.
- Submit pictures of volunteers to be in your organization's newsletter.
- Post pictures of volunteers on a bulletin board at your organization.
- Provide organizational goodies – hats, shirts, pins, magnets, water bottles, etc.
- Have them join you for coffee or lunch.

Below are some more involved, intermediate recognition ideas:

- Nominate a volunteer Star of the Month – award them a certificate, letter, or small gift.
- Sponsor happy hours and social events. Encourage volunteers to meet each other.
- Recognize volunteers on local radio or television stations.
- Invite volunteers to serve as project leaders or committee members.
- Give gift certificates to museums, movies, restaurants, etc. Solicit your community for donations!
- Nominate volunteers for local/national awards such as the Presidential Service Awards.
- Write articles about them in newsletters or newspapers.
- Write a letter to their employer highlighting the accomplishments of the volunteer. Be sure to find out if the volunteer would appreciate this before writing the letter!
- Celebrate major accomplishments.
- Recognize anniversaries with your organization.
- Have them attend a training, workshop, seminar, etc. at the expense of your organization.
- Give them additional responsibilities.
- Create a photo collage or slide show of volunteer activities.

Some large-scale means of recognition:

- Hold annual recognition events: a dinner, a breakfast, an awards ceremony/celebration, a picnic/potluck, theme party, etc.
- Recognize long-term volunteers with Service Awards: a plaque, trophy, certificate, etc.
- Give additional responsibilities and a new title.
- Put up a banner celebrating major accomplishments.
- Enlist them in training staff and other volunteers.
- Involve them in the annual planning process.
- Make a donation to the organization of their choice in their name.
- Organize an outing at a zoo, amusement park, sporting event, etc., where volunteers get in for free.

Recognition is an important part of volunteer management. Recognition is an opportunity for the community, other volunteers, and program staff to show their appreciation for the volunteers' efforts. When tailored to meet the volunteers' needs, recognition helps them feel good about themselves and their service.

The above information is a compilation of materials from Make A Difference, a 501(c)(3); *Volunteer Management* by Steve McCurley and Rick Lynch (1997); http://www.energizeinc.com; and http://www.casanet.org.

SUMMARY

Without the right number of volunteers with the right skills, your service project won't be successful. Whether recruiting volunteers for a one-day service project or for a weekly commitment, you should understand your volunteer needs and then target recruitment efforts to reach the volunteers you want to engage.

Offer opportunities that will appeal to their interests and work with their schedules. Retain volunteers by supporting them before, during, and after the project. Finally, recognize their efforts in a way that makes them feel appreciated and connected to the community.

From: Corporation for National & Community Service & Hands On Network, "Take Root: Volunteer Management Guidebook," (2010).

13.3 The Hierarchy of Ethical Values in Nonprofit Organizations:

A Framework for an Ethical, Self-Actualized Organization Culture

By Ruth Ann Strickland and Shannon K. Vaughan

OBJECTIVES:

10. Describe each of the five levels in the ethical hierarchy of behavior for organizations.
11. Define the term "integrity" as used for the highest level in the ethical hierarchy.
12. Name examples of nonprofit agencies that have suffered ethical scandals.
13. List external controls that can force nonprofit agencies to act ethically.
14. Identify actions agencies can take to achieve respect for volunteers and employees.

ABSTRACT

Using Maslow's theory of human psychological development as a framework, a model based on the hierarchy of values is proposed to explain how not-for-profit organizations develop an ethical culture. As with individual values, the five levels of ethical behavior—financial competence, accountability, reciprocity, respect, integrity—are attained successively and one at a time. Thus ethical values are a foundation for achieving integrity, defined herein not only as incorruptibility but as a total commitment to the highest standards of behavior. External controls stimulate ethical behavior primarily at the lower levels; internal controls must be present to achieve an ethical organizational culture.

Inherent in the concept of ethics is a recognition of right and wrong in the decision-making behavior in an organization. According to Hansmann's (1980) theory of contract failure, nonprofit organizations are often the venue of choice for service delivery because they are deemed more trustworthy than business. While this is true for many reasons, numerous scandals involving nonprofit organizations have illustrated that the third sector is not immune from ethics problems. What, then, makes a nonprofit organization ethical? Studies have shown that organizational culture is one of the most important factors, if not the most important, influencing ethical behavior, especially with regard to integrity (Hendershott, Drinan, and Cross 2000). To enhance the understanding of nonprofit governance, this article proposes a model based on Maslow's hierarchy of needs for the development of an ethical culture within a not-for-profit organization. First published in 1943, Abraham Maslow's *A Theory of Human Motivation* relies upon a hierarchy of needs to explain why individuals are motivated to act. Maslow sets forth five levels of needs in a firmly hierarchical structure, with the satisfaction of lower-level needs prerequisite to the attainment of the next-higher level. For example, until individuals satisfy their basic physiological needs for food, clothing, and shelter (Maslow's first level), they will be less likely to recognize opportunities to meet higher-level needs, such as love and self-esteem, or expend the effort to meet them. Self actualization—defined by Maslow as "what a man *can* be, he *must* be" (1943, 382; emphasis in original)—is the pinnacle of the hierarchy; self-fulfillment is the ultimate motivation to act, but lower-level needs that go unmet prevent its attainment.

Whereas some critics suggest that Maslow's pyramid of needs is not really a hierarchy but instead has cyclical properties, Maslow clearly theorized that the levels of needs have a scalar quality. One does not start over satisfying physiological needs once self-fulfillment is attained. Rather, because satisfaction of each level of needs is not finite—once achieved, they are not simply crossed off the list never to be faced again—scaling the hierarchy is more like climbing a mountain than completing a cycle. Climbers do not reach the summit of Mount Everest without several strategic detours back and forth to the lower levels of the mountain, but each time the detour down becomes easier and less resource-consuming. Likewise, central to the understanding of Maslow's theory of motivation is that satisfaction of lower-level needs gives individuals the slack resources to focus on a larger goal; needs that are consistently unmet divert attention from pursuing little beyond their satisfaction.

Maslow (1943) formulated an enduring and provocative theory of human motivation. Scholars have both venerated and disparaged his theory, but it is continually cited and tested more than six decades after his initial contention that individuals act based upon a hierarchy of needs. Not only does Maslow's work have mainstream appeal—with references in the popular press ranging from the *Wall Street Journal* and *Forbes* to *Sports Illustrated*—but his hierarchy is utilized extensively in scholarly work. Research employing the theory has been published in numerous academic journals, including *Engineering Management Journal, Operations Research*, the *Journal of Research in Personality*, and *Public Administration Review*. The hierarchy of needs has been applied to many different topics, including national development (Bailey 2005), business ethics (Hatwick 1986), organizational behavior (Cullen and Gotell 2002), motivation (Atwood 2004; Borkowski 2005; Halepota 2005; Janiszewski 2005; Rouse 2004), organizational resource allocation (Ivashchenko and Novikov 2006), information technology management (Coffee 2002; Pisello 2003), dispute resolution (Duffy and Thomson 1992), and terrorism (Schwing 2002). This body of literature gives testimony to the multidisciplinary applicability of the approach, as do the numerous introductory psychology, business management, and public administration textbooks that include discussion of the hierarchy of needs as a critical element in the study of motivation.

Although Maslow's theory is not always supported in the vast literature to which it is applied and may not apply cross-culturally, it remains an enduring framework for examining human and organizational behavior. Accordingly, the hierarchy of needs is employed herein as the basis for developing a new model for understanding the ethical behavior or lack thereof of not-for-profit organizations. As with Maslow's hierarchy of needs, it is expected that the hierarchy of ethical values in nonprofit organizations will post interesting challenges for those who seek to test it empirically, and that the challenges will either be overcome in similar fashion or provoke a stimulating debate (Latham and Pinder 2005; Lord 2002).

The discussion begins by introducing the hierarchy of ethical values. Each level is discussed in relation to Maslow's hierarchy of needs, but the model focuses on organizations and their ethical development, not on the fulfillment of individual needs within organizations. After a description of the model, internal and external controls will be discussed in relation to achieving each level in the model. Examples are provided as illustrations of the theoretical basis of the model (not as empirical tests) and serve to highlight how nonprofits exist in various stages of ethical development. Finally, the implications for the voluntary sector when nonprofits fail to seek the highest level of integrity as well as the need for further research in this area are examined.

HIERARCHY OF ETHICAL VALUES IN A NONPROFIT SETTING

To create and internalize ethical behavior, an organization must first attain certain basic ethical values. Attributes such as financial competence and accountability (Levels 1 and 2, respectively) must be attained before an organization can hope to achieve integrity (Level 5). Outside efforts to establish ethical boundaries are commendable, but the real work of creating an ethical organizational culture resides within the organization itself.

It is important to internalize an organizational culture that embraces key ethical ideals and procedures and makes them central to the organization's everyday operations (Jeavons 2005, 206). Organizations that lack an ethical compass inevitably damage their own interests as well as donor interests and may indirectly harm all others in the third sector (Schmidt 2004). The ethical hierarchy of values serves as the framework for fostering an ethical culture by encouraging (1) financial competence (managing resources and assets wisely), (2) accountability (transparency), (3) reciprocity (maintaining a mutually beneficial investment relationship with donors to meet the needs of targeted constituencies), (4) respect (incorporating the perspectives of employees, volunteers, and donors into all organizational activities), and (5) integrity (preserving incorruptibility and completeness in commitment to the mission). These values are the foundations for creating an ethical organizational culture and environment, as shown in Figure 1.

Figure 1. Hierarchy of Ethical Values for Nonprofit Organizations

Level 1: Maintaining Competent Financial Management

Managing assets wisely and maintaining solvency is equivalent to Maslow's concept of individual physiological needs. This value sets the organization up for success in all other areas. Financial competence promotes successful budgeting and recruitment/ retention of staff, volunteers, donors, and clients. Not-for-profits that lack the proficiency or commitment to safeguard financial resources or that use unscrupulous methods to corral resources fail to exhibit the most basic level of ethical behavior. Without achieving this most basic level, nonprofits are stymied in their efforts to articulate and pursue their mission.

As the number of not-for-profit organizations has increased dramatically in recent years, so has the competition among them for resources. Chasing financial resources, unfortunately, can lead to mission drift (Grace 2006) or vendorism (Salamon 1995). At worst, lack of fiscal competence leads to financial mismanagement, as exemplified by inaccurate financial reporting, excessive executive compensation packages, misuse of donor monies, and poor auditing procedures (Weiner 2003, 56). Lack of competent financial management carries a high price, not just in monies lost, but also for the organization's clients and for donors who put their trust in the ability of the nonprofit to achieve its stated mission. Like individuals who cannot pursue higher-level needs when they are physiologically deprived of food, water, or shelter, organizations that are financially insolvent or mismanaged lose sight of the mission.

Level 2: Establishing Accountability

Accountability refers to the ability of nonprofit organizations to establish transparency and trust. It is equivalent to Maslow's concept of individual safety needs; accountability equates to safety in this regard because nonprofit organizations are more likely to attain security if they set up transparent procedures as well as proper oversight. When organizations follow external controls and even conform to higher standards of accountability, they are less vulnerable to scandal. In addition, accountability equates to safety in that nonprofits that establish accountability have taken steps to protect themselves from unethical behavior, thereby preventing the expenditure of resources to investigate or mitigate the consequences of such behavior.

To establish accountability, nonprofits must ask themselves who they are accountable to, for what, and how. They are held accountable internally by their own board's governance procedures; they are held accountable externally by the Internal Revenue Service and other government regulators. In addition, nonprofits also must be responsive to their stakeholders (donors, staff, members, clients, contract managers, and volunteers) as they pursue mission and maintain program effectiveness. Finally, the general public represents the broadest category of stakeholder. Since nonprofits benefit from tax expenditures as well as direct funding by government, taxpayers and citizens have a right to monitor nonprofit activity and its value to society (Brody 2002, 473). Because nonprofits often face multiple, sometimes conflicting demands from a wide array of stakeholders, defining accountability in a way in which one size fits all is not appropriate. Still, nonprofit organizations achieve accountability only by keeping the questions of who, for what, and how foremost in their service to constituencies.

Level 3: Establishing Reciprocity

Equivalent to Maslow's concept of individual affiliation needs, reciprocity refers to the ability of nonprofit organizations to serve their constituents and donors in a manner that maximizes acceptance and trust. In practice, ethical fundraising is an example of how organizations can meet reciprocity needs. While nonprofit organizations rely on contributions to function, Grace (2006) argues that they should move beyond the beggar's tin cup and focus on match. Matching a donor's interests with a nonprofit's needs is analogous to Maslow's level of affiliation. Pursuing donors who share the organization's mission yields mutual benefits. Without reciprocity, nonprofit organizations may experience goal displacement and diverge from their mission as they pursue funds rather than partners.

By definition, not-for-profit organizations do not distribute excess revenues to third parties but retain them within the organization; therefore, they are not about making money but about providing services. As such, mission is the driving force of a nonprofit organization. It is important, therefore, that nonprofits have and adhere to a clearly articulated statement of mission and purpose (Werther and Berman 2001; Wymer, Knowles, and Gomes 2006). Board members, staff, and volunteers need a clear understanding of what the organization seeks to do, how, and why. Because the third sector depends heavily on goodwill and trust, adherence to a clear mission statement enables nonprofits to be better accountable to their supporters, members, clients, donors, and the public by making it clear how they will fulfill their philanthropic goals (Jeavons 2005, 218).

Healthy nonprofit organizations place importance on the

specific interests of the donors, volunteers, and staff associated with them, thereby employing a supplyside rationale. Supply-side functions also include social entrepreneurship, whereby commercial venues are used to foster charitable goals (Frumkin 2002; Young and Salamon 2002). Nonprofits achieve reciprocity when they recognize and celebrate the match between donor interests and their own.

Level 4: Instilling the Value of Respect

The ability to attain status and respect is important to a nonprofit organization's credibility and is equivalent to Maslow's concept of individual esteem needs. At this level of organizational culture development, the nonprofit is respected by others and, as well, has respect for itself and others. Employees feel worthwhile and appreciated. Donors, volunteers, and clients are treated as integral members of the team.

What Grace (2006) terms the donor-investor relationship embodies how nonprofits garner respect. Grace implores nonprofits to take a development rather than fundraising approach to resource recruitment. Development involves cultivating relationships with donors that induce them to view their contributions as an investment in the work being done by the nonprofit organization.

Developing relationships with donor-investors goes beyond simply asking for money. Donors are viewed as integral team members, with a specific interest in the work being done and a desire to invest in the organization as a whole, not simply to write a check. Because philanthropy is defined as "all voluntary action for the public good" (Grace 2006, 1), volunteers are celebrated as donor-investors.

Level 5: Integrity and the Self-Actualized Nonprofit Organization

The highest value in an ethical organizational culture is integrity, equivalent to Maslow's concept of individual self-actualization. Integrity is defined not only as incorruptibility, but as completeness of commitment to ethical behavior. With integrity, an organization has an internalized moral code, is able to engage in creative problem-solving, and pursues its mission to the fullest extent possible. Nonprofits that have achieved integrity assume a stewardship role in serving the public.

Aspiring to integrity and fulfilling the ethical hierarchy of needs is important if nonprofit organizations are to enjoy the full confidence of the public. Ethical governance of nonprofits is necessary to maintain their integrity. Attaining integrity relies on achieving financial competence, accountability, reciprocity, and respect. Building and maintaining social capital is essential to the ability of nonprofit organizations to mobilize support and engage in collective action (Jeavons 2005, 223).

Scaling the Hierarchy

Can ethics be regulated? This framework identifies the levels of ethical development whereby nonprofit organizations reach and attain integrity. While nonprofits can be taught ethical practices, they cannot be forced to act ethically. External controls can be imposed on nonprofits to move them through financial competence and accountability, as well as to contribute to achieving reciprocity. However, only by internalizing ethical behaviors and patterns can a nonprofit attain integrity.

Internal and External Controls

Legislation, Internal Revenue Service (IRS) rules and regulations, and contract stipulations for the receipt of grant funds are all external controls on the operations of a nonprofit organization. Often these requirements are precipitated by scandals that raise awareness of a particular vulnerability. External controls may be sufficient to impose financial competence and accountability on a nonprofit, but without an internalized commitment to ethics, the organization will not move beyond Level 3—Reciprocity.

The following sections present examples of scandals that have affected nonprofit organizations. The examples should not be interpreted as empirical evidence but rather as descriptors of the model's concepts to facilitate empirical tests. They illustrate the obstacles to moving through the hierarchy of ethical values and prescriptions for overcoming them. Although the focus here is on external controls as both necessary and sufficient to achieve Levels 1 and 2, attention is also paid to the internal control mechanism that is crucial for organizations to reach the upper levels.

Financial Mismanagement (Level 1)

Like all organizations, nonprofits are not immune to scandal. Allegations of financial misconduct are the most prevalent, and ultimately the most damaging to the nonprofit organization and the voluntary sector as a whole. Charged and subsequently convicted of fraud and misuse of donor funds, Jim Bakker caused a scandal that not only brought about the demise of PTL Ministries in 1987 but contaminated other evangelical ministries by harming their reputations and their fundraising efforts (Jeavons 2005, 214). The Ohio division of the American Cancer Society suffered stinging repercussions from the loss of $7 million through embezzlement. Proper safeguards were not in place to prevent this—the same employee kept records, reconciled bank accounts, and had direct access to organization funds. The questions raised concerned not only why this individual had such easy access, but also why the organization had $7 million cash on hand ("Theft from Cancer Society" 2000).

A Chronicle of Philanthropy review of 10,770 nonprofit organizational tax records from 1998 to 2001 revealed that more than 1,002 charities made $142 million in loans to their directors, officers, or key employees (Davis 2004, 1). In 2004, People United for a Better Oakland (PUEBLO) came under fire for routinely making personal loans to board members and the executive director. The board chair claimed that organizations that work with poverty often have to take such "emergency measures" (Jackson and Fogarty 2005, 125). However, the loans were not made to the poor, unemployed, or disenfranchised; rather, they were made to (and sometimes not repaid by) board members, employees, and organization supporters. In addition, about $500,000 was unaccounted for between March 2002 and March 2004 (Johnson 2004).

While staff members of nonprofits are typically underpaid relative to the market, some large nonprofit organizations often argue that it takes high salaries to attract capable executives to assist in fundraising and attract major gifts. On June 14, 2004, Carl Yeckel, former president of the Dallas-based Carl B. and Florence E. King Foundation, and Thomas Vett, the foundation's former secretary, were ordered by a jury to pay $14 million in compensatory and punitive damages to the King Foundation. Yeckel and Vett were censured for excessive executive salaries and amassing personal charges on the foundation's credit cards (Osborn 2004).

In 2001, Hale House (a nonprofit dedicated to serving drug-addicted and abandoned babies in Harlem) faced scandal when shelter director Lorraine Hale was accused of stealing money from the organization. Hale and her husband were later sentenced to five years of probation and ordered to pay restitution for the $766,000 they had embezzled (Saltonstall and Evans 2004, 32; "Shelter's Ex-Director" 2002). After the scandal broke, the number of donors dropped dramatically from 200,000 to 12,000; two years later, the donor base had increased to only 50,000, one-fourth the pre-scandal size (Souccar 2004, 14).

Each of the scandals mentioned above involved issues of financial competency. Early in their development, nonprofit organizations may have more lax fiscal systems in place, due either to the administrative inexperience of the leadership or to a high degree of assumed trustworthiness within the group. Financial mismanagement is less likely to occur in organizations that internalize ethics early; for those without a strong internal orientation toward ethics, disasters like the ones described above may ensue.

Accountability (Level 2)

The prohibition against distributing profits means, in theory, that members of a board of directors have no other incentive than to act according to the best interests of the organization and its clients. Trust as a substitute for monitoring, however, is problematic when it leads to a disengaged board that is more susceptible to scandal. Transparency facilitates engagement and is an antidote to scandal.

Two members of the board of directors of the United Way of the National Capital Area (UWNCA) were removed when they pressed for open financial records; they pressed for access after being told they were not entitled to see financial statements (Strom 2003, 1). Subsequently, a top executive stole $500,000 from the charity and its pension fund (Hananel 2004). Later, the entire board of directors was replaced after allegedly inflating the organization's fundraising figures, understating overhead costs, and overcharging for administrative fees (Owen 2003).

The Nature Conservancy came under heavy scrutiny after the *Washington Post* reported that it had purchased land from Georgia-Pacific during the time that Georgia- Pacific's chairman sat on the Nature Conservancy board (Bobelian 2004, 4). Conservancy board members sold land to the Conservancy and then bought property from it. The leadership of the Nature Conservancy was also roundly criticized for not more carefully scrutinizing tax deductions taken by donors and for failure to make its finances more public (Stephens 2004, A01).

In 2004, a local California chapter of the American Red Cross released detailed reports on how it had spent funds after the wildfires in October 2003. The reports revealed that the nonprofit had spent 67 percent (or $3.9 million) of its funds directly on fire victims. This stands in stark contrast to revelations from the Alpine fire in 2001, when an audit showed that only 10 percent of funds raised went to the fire victims (Vigil 2004). A scandal ensued in which fire victims and the public wanted to know how donations were spent and how funds were managed.

Achieving Levels 1 and 2

Although organizations cannot be forced to act ethically, legal requirements can encourage achievement of Levels 1 and 2 of the hierarchy. Organizations receive (and maintain) tax-exempt status from the Internal Revenue Service only after meeting legislatively established criteria. Therefore, nonprofits must meet financial competence and accountability requirements, such as filing IRS form 990, to continue to benefit from tax-exempt status.

More than half of all tax-exempt organizations are classified as public charities under section 501(c)(3) of the Internal Revenue Code (Boris 1999). Public charities are subject to greater scrutiny by the IRS because they are afforded the added privilege of tax-deductibility of all contributions made to them. Therefore, the IRS has stricter criteria for recognition as a public charity and for maintaining that status (IRS 2006).

In response to the scandals involving Enron and other companies, Congress passed the Sarbanes-Oxley Act (SOX) in 2002 to deter fraud (Ostrower and Bobowick 2006). Two provisions of the act apply to all organizations, including nonprofits. Although whistleblower protection and document-retention criteria are the only mandates that apply directly to nonprofits, the act contains several other provisions that have been recognized as best practices for nonprofit governance. California's Nonprofit Integrity Act requires implementation of some of the best practices from SOX by nonprofits in that state (Jackson and Fogarty 2005).

One of the SOX best practices provisions involves auditing committees. Audit committees are a conduit between the board and the outside auditor, enhancing communication and information flow. By ensuring that the organization meets its financial responsibilities and disclosure requirements, the audit committee is positioned to identify financial irregularities before they become problematic (Owen 2003).

The burden of complying with the enhanced auditing provisions of SOX depends on the provision itself. Many nonprofits already comply with some provisions, whereas others would find it very difficult to enact the provisions. More than half of the 5,115 nonprofits surveyed in the Urban Institute's National Survey of Nonprofit Governance stated that it would be somewhat or very difficult to comply with the provisions for establishing an audit committee. More than two-thirds said it would be difficult to comply with the requirements to rotate audit firms or lead auditors (Ostrower and Bobowick 2006).

The Sarbanes-Oxley Act also includes deterrence measures regarding conflicts of interest in publicly traded companies. Extending these provisions to nonprofit organizations would enhance accountability by facilitating greater transparency regarding board members' activities/interests. Conflict of interest may occur when a board member makes a decision out of self-interest or in the interest of only part of the organization; conflicts also can occur when the nonprofit does business with or has a financial link to a board member or a relative of the member. Internal controls regarding conflict of interest involve recusal from the decision-making process when a potential conflict of interest exists (Soltz 1997, 131), as well as development and periodic review of conflict of interest policies (Tyler and Biggs 2004, 22). According to the Urban Institute survey, 50 percent of nonprofits have a conflict of interest policy for their board members. This may be misleading, however, because while 95 percent of large organizations have such policies, only 23 percent of small ones do (Ostrower and Bobowick 2006). This suggests that resource capacity may be an issue; small organizations are more likely to be focused on issues of financial competence and accountability, with fewer slack resources to devote to concerns about conflict of interest. Therefore, organizations focused simply on meeting the external control requirements associated with Levels 1 and 2 are less likely to pursue internal controls, such as a conflict of interest policy.

Investigations by state attorneys general, auditors, or other officials also provide external controls and prompt nonprofits to adopt their own internal controls. In 2002, Ohio auditor Jim Petro found that Specialized Alternatives for Families and Youth of Ohio Inc. (SAFY) misspent state funds, using the money to buy new businesses rather than putting the funds toward the care of children. SAFY made changes in response to the audit by installing a new board of directors, new accounting software, and new policies as recommended by Petro (Bischoff 2002).

Watchdog groups also provide an external check. Some watchdog groups examine the spending practices of nonprofits, reporting the ratio of funds spent for administrative costs relative to program activities. These groups also make statements regarding the degree to which legal activities are actually ethical when practiced by nonprofits. For example, the practice of insider loans is legal, but as the example above demonstrates, this activity is not always ethical when donor funds are involved. Board members may find themselves personally liable if insider loans are not repaid (Franklin 2004). Unless a real benefit accrues to the organization as a result of the loan, private loans could jeopardize a nonprofit's tax-exempt status as well as its legitimacy with donors, thus threatening organizational security. Insider loans, if used at all, should stipulate a short-term loan at a market interest rate, with repayment closely monitored.

The American Institute of Philanthropy (2003) advocates

reform measures to make boards more independent, more engaged in oversight, and more aware of the implications of their decisions. Board members should receive training on how to interpret financial reports, how to exercise oversight of budgetary matters, how to make decisions on employee and executive compensation, and how to treat staff and volunteers. In addition, state attorneys general are empowered to enforce board duties of care and loyalty (Brody 2002).

In order to ensure greater accountability to donors, some nonprofit advisory groups propose that nonprofits adopt a donor's bill of rights. The ten recommendations require that organizations disclose how they will use donated funds, provide the identities of the board members, and share their most recent financial statements. The bill of rights also requires nonprofits to assure donors that their donations will be used for the stated purposes for which they were given, to properly recognize donors, and to ensure that donations are handled with confidentiality. Finally, donors should be informed of whether those seeking donations are volunteers, staff, or hired marketers, and they should feel free to ask questions and receive forthright answers when making donations (Watson 2000).

The Johns Hopkins Nonprofit Listening Post Project—a March 2005 survey of 443 organizations with 207 respondents—found that 93 percent of nonprofits distribute financial statements to their boards on a quarterly basis, and 62 percent share them every month. Seventy-four percent make their financial reports available to members of the public upon request, 70 percent distribute them to donors, and 54 percent publish their statements in annual reports. Nine percent post financial reports on their organizational Web sites (Salamon 2005). These results indicate that a great many nonprofit organizations are committed to achieving accountability, by means in excess of what is required by law. Such internal commitment to ethical values bodes well for the achievement of integrity in nonprofits.

Achieving Reciprocity (Level 3)

A match between donor interests and the nonprofit's mission is critical to achieving reciprocity. As nonprofits evolve, they become more professional in orientation and outlook. It follows, therefore, that they will become more attuned to their mission and their relationship to the community as a whole. Chasing funding sources without a clear connection to mission weakens the organization, impedes reciprocity, and opens the organization to scandal.

In 2003, the Kids Wish Network—a nonprofit established to offer comfort and hope to children with terminal or life-threatening illnesses—collected $205,255 in donations in New York through the work of professional fundraisers. After the fundraisers were paid, a mere 12 percent ($24,634) was retained by Kids Wish Network. In a spot check of 607 fundraising campaigns in 2002, the attorney general of New York found eight other professional fundraisers that turned over a meager 12 percent of proceeds to the charitable organization for which they were raised (Gormley 2003).

The American Institute of Philanthropy and the Better Business Bureau recommend that charities keep at least 65 percent of the monies raised by the professionals. A law enacted in California institutes more protections for consumers, making nonprofits more accountable for hiring fundraisers (Gormley 2004). For example, the Maryland Association of Nonprofits recommends that organizations work to ensure that over a period of about five years, on average every dollar spent on fundraising should be matched by raising at least three dollars (Causer 2004; Salmon 2004).

Many believe that making charities profitable for fundraisers erodes the principle of reciprocity—that is, that nonprofits should be responsive to donors and designated constituencies. External controls in this area are limited because the U.S. Supreme Court has consistently invalidated state laws that place numerical limits on fundraising ratios on the grounds that such limits are too restrictive of free speech and association. Most states provide ethical guidelines and publicize fundraising ratios (Bryce 2005), but achieving reciprocity ultimately requires internal control. Nonprofits that employ Grace's (2006) development approach to resource attainment are more likely to achieve reciprocity by cultivating donor-investors who support and contribute to the mission of the organization.

Although not-for-profits are prohibited from using federal grant or contract funds for lobbying activity, Congress and the Internal Revenue Service (IRS) generally support advocacy activities (including lobbying with private funds) by nonprofits. The Tax Reform Act of 1976 clarified and expanded the scope of lobbying activity permissible by 501(c)(3) organizations, specifically by narrowing the legal definition of lobbying subject to restriction. Lobbying is differentiated from other advocacy activity because it occurs only when there is an expenditure of funds by the not-for-profit organization for activities aimed specifically at influencing legislation. Advocacy involves providing information in an effort to educate about and promote an issue or overall policy response ("Charity Lobbying" in the Public Interest, n.d.;

Smucker, 1999).

The Internal Revenue Code states that a 501(c)(3) organization "may not attempt to influence legislation as a substantial part of its activities and it may not participate at all in campaign activity for or against political candidates" (IRS n.d.). Questions have arisen, however, regarding what constitutes less than a substantial amount of time that nonprofits can legally spend in lobbying activities and at what point these organizations become too political. In 1990, in response to the confusion over how much is too much time spent lobbying, the IRS implemented the expenditure test—also known as the H election—as an alternative to the substantial-part rule inherent in the relevant section of the Internal Revenue Code. Nonprofits must elect to fall under the provisions of section 501(h), which prescribes specific limits on lobbying expenditures and types of lobbying, and sets some protections for organizations that commit single-year violations (Smucker 1999).

Whereas it is possible to institute external controls that facilitate reciprocity, such as reporting requirements for receipt of funds and legal restrictions on the use of funds for lobbying activity, fully achieving reciprocity requires an internal commitment. The development of donor-investors requires an internalization of an ethical commitment to serving constituents and celebrating those who contribute to the nonprofit's efforts. A match between donor interests and nonprofit mission is of paramount importance.

Achieving Respect (Level 4)

Just as a match between donor interests and organization mission is important, a match between staff and volunteer interests is likewise essential for developing an ethical culture. Recruiting and retaining the most appropriate board members, staff, and volunteers is critical to fulfilling the nonprofit's mission. Each individual should be encouraged to engage in dialogue about his or her perceptions of the organization's mission. Tolerance of differences of opinion and cultural diversity not only make nonprofits successful, it creates a respectful organizational culture (Watson and Abzug 2005, 628).

Proper training, assignment, risk management, and motivation are crucial elements in respecting everyone associated with a nonprofit. Nonprofit organizations often fail to provide proper training for staff members who supervise volunteers, assuming that these skills come naturally. This can result in ethical and legal dilemmas. Without training, staff may be unaware of many human resource management pitfalls. Staff members who supervise volunteers should be trained on performance-appraisal, motivational, and recognition techniques (Macduff 2005, 715).

Clear job descriptions that explain the work needed, the skills required, the tasks involved, and supervisory arrangements establish boundaries for how volunteers will be used in the organization. Job descriptions enhance volunteer proficiency, and a clear outline of responsibilities prevents going into areas beyond skill level. Volunteers are more likely to feel good about their work and motivated to continue if they are equipped to be effective (McCurley 2005, 607–608). In addition, volunteers should not be asked to do work that paid staff would never be asked to do. If asked to do work that paid staff perform, volunteers should receive commensurate training (Bradner 1997, 171). Finally, organizations have an ethical obligation to provide liability insurance for volunteers (Brudney 1999, 241). By reducing volunteer fears of liability and properly training them to stay within the scope of their responsibilities, not only does the organization protect itself and its volunteers, it also better serves its clients.

Protecting staff is also important, especially staff members who seek to expose wrongdoing within the organization. Although the Sarbanes-Oxley Act prohibits retaliation against whistleblowers who provide truthful information related to the possible or actual commission of a federal offense, many nonprofits either have not formulated adequate protections for whistleblowers or are behind in implementation. A 2003 survey of 300 nonprofit CEOs revealed that 57 percent are familiar with SOX, and of these, 80 percent head nonprofits with $10 million in revenues. Yet only 20 percent of the CEOs had changed their governance policies to comply with SOX (Sinclair 2004).

For example, Dulcy Hooper, who worked for the United Way of the National Capital Area (UWNCA), told her superiors about inconsistencies in gift reports. Not long after sharing her concerns, she was labeled "not a team player" and lost her job. She was one of many whistleblowers who were shunted aside and characterized as troublemakers. A forensic audit later conducted on UWNCA revealed years of financial mismanagement. Perhaps if the organization had listened to the natural, built-in early warning system of their gifts officer, it could have avoided a great deal of adverse publicity (Sinclair 2004).

SOX's protections for whistleblowers provide an external control over organizational behavior, but they constitute after-the-fact enforcement as compared to the cultivation of ethical culture. Protection of individual staff members is imposed by prohibiting retaliation against whistleblowers, but respect for the same staff members is only facilitated when their comments and concerns are given legitimate concern prior to the need to blow the whistle. An ethical culture within a nonprofit organization means that employees are encouraged to blow the whistle, not merely protected once they have done so. External control, therefore, is not sufficient to embody respect and achieve Level 4 within a nonprofit organizational culture.

Achieving Integrity (Level 5)

Identifying examples of nonprofits that have achieved integrity is more challenging than identifying those that achieve financial competence, accountability, and reciprocity. Because external controls are more applicable to ensuring these types of ethical behaviors by organizations, monitoring of their successful accomplishment is more feasible. Also, as with other issues, bad behavior makes the news, whereas good behavior usually does not. Numerous watchdog groups, such as Charity Navigator and the American Institute of Philanthropy, have developed rating scales of top nonprofits, evaluating organizations based primarily on financial competence, transparency, and protection of donor interests. The nonprofits that consistently rate high on these elements are the ones most likely to achieve integrity (based on the hierarchy of ethical values proposed herein). These ratings may provide a good starting point for identifying the level of ethical culture development in nonprofit organizations.

Boys & Girls Clubs of America, the Nature Conservancy, the Mayo Clinic, and the American Red Cross are well known, and their names are brands. Branding represents a promise of organization principles, operational values, and the benefit the organization seeks to deliver to society (Wymer et al. 2006). Staff, volunteers, donors, clients, and the general public feel a sense of pride in what the organization has accomplished and, more important, trust the means by which the organization conducts its work. In addition, each of these organizations made Charity Navigator's top-ten list of the "Best Charities Everyone's Heard Of" (Charity Navigator 2006).

Achieving Level 5 means possessing a brand of integrity whereby relevant stakeholders and the general public believe that the organization has fulfilled the elements at each of the lower levels—financial competence, transparency of operations, affiliation/alliances, and genuine respect for everyone involved with the organization's work. Nonprofit organizations of integrity exhibit a stewardship approach to management and administration. According to stewardship theory, stewards place higher value and priority on collectivist rather than individualistic behaviors, that is, on cooperation rather than defection (Davis, Schoorman, and Donaldson 1997). Because the organization's performance is the primary focus, stewards are able to maximize the satisfaction of all stakeholders (including the steward's) by acting in the organization's best interests.

An ethical organizational culture in the nonprofit sector is essential to stewardship. Stewards must be vested with a high degree of trust, and therefore an internalized ethical culture is crucial to develop staff, volunteers, and board members into stewards. External controls, such as legal mandates for reporting, rules and regulations regarding financial management, independent watchdog groups, and so on, can only influence ethical behavior to a certain degree. Unless the individuals within the nonprofit work to ensure that the culture of the organization facilitates ethical conduct, integrity will not be achieved. For example, McCabe and Trevino (1996, 29) suggest that the key to curbing cheating in academia may be to "create an environment where academic dishonesty is socially unacceptable." Disapproval of cheating among peers is a chief determinant of whether students change their cheating habits between high school and college (Hendershott et al. 2000).

The importance of culture is also reflected in what Frumkin (2002) terms the expressive rationale, whereby nonprofits exist due to the desire of stakeholders to express their values and faith. The concept of stewardship is probably the most prevalent among faith-based and environmental nonprofits. For example, the National Christian Foundation—number 1 on Charity Navigator's list of "10 Best Charities Everyone's Heard Of"—defines faithful stewards as "people who understand what they hold belongs to God" (NCF 2006). Likewise, Conservation International (number 5 on the list) "believes that Earth's natural heritage must be maintained if future generations are to thrive spiritually, culturally and economically" (CI 2006). Each of these statements implies placing individual interests secondary to the community (and organization) as a whole.

CONCLUSION: IMPLICATIONS OF NOT STRIVING FOR INTEGRITY

Simply following the letter of the law does not mean that an organization is ethical; external controls can only take an organization partway to developing an ethical culture. Many nonprofits caught up in scandal broke no laws. However, sexual misconduct by staff, excessive compensation packages for executives, drift from the organization's mission, and questionable fundraising practices all erode public confidence in the nonprofit sector. Internalization of ethics through the development of an ethical culture is necessary to ensure the integrity of the nonprofit sector so that charitable organizations can thrive.

Nonprofit organizations, like individuals, usually do not start out exhibiting the highest level of ethical behavior. Just like other organisms, nonprofits evolve, from formation through growth to maturity (Werther and Berman 2001). Likewise, nonprofits will move through the hierarchy of ethical values in a series of stages. External controls are most relevant at the first two levels of the hierarchy; laws, rules, and regulations provide useful structure and guidelines during the early years as nonprofits form their organizational culture. Reciprocity is a level of transition, as external controls become less important than internal controls in shaping the ethical behavior of the nonprofit. Respect and integrity are values achieved only through an internalized ethical culture throughout the organization.

The framework articulated in this article is intended to foster understanding of the ethical behavior or lack thereof in nonprofit organizations. The natural next step is to test the model. As mentioned, Maslow's hierarchy has been subjected to extensive study with varying results. Because the concepts built into the model are inherently subjective, empirical testing may be difficult, but it still is possible. This work is important; understanding what drives the ethical behavior of nonprofit organizations is especially relevant given the dramatic growth in the third sector and the increasing attention paid to ethical conduct given recent scandals across all sectors—nonprofit, private, and public.

The costs of engaging in unethical conduct far outweigh the benefits. Meeting ethical values, such as financial competency, accountability, reciprocity, and respect, empowers nonprofit organizations to fulfill their missions and to retain public trust and confidence. The long-term success of the voluntary sector will only be possible if nonprofit organizations internalize these values and become self-actualized.

References

American Institute of Philanthropy. 2003. "Are Charity Boards Asleep at the Wheel? Nonprofit Governance Problems" (www.charitywatch.org/articles/asleep.html).

Atwood, Jeffrey. 2004. "Employee Motivation of the High Achiever." SuperVision 65, no. 5:3-4.

Bailey, Charles R. 2005. Winning the Hearts and Minds: Providing the Basic Needs First. Carlisle Barracks, Pa.: U.S. Army War College.

Bischoff, Laura A. 2002. "State Audit Challenges Foster Care Agency's Costs." Dayton Daily News, October 10:B5.

Bobelian, Michael. 2004. "Following the Nonprofits." Recorder (June 1):4 (http://web.lexis-nexis.com/universe).

Boris, Elizabeth T. 1999. "Nonprofit Organizations in a Democracy: Varied Roles and Responsibilities." In Nonprofits and Government, edited by Elizabeth T. and C.

Eugene Steuerle, pp. 3-29. Washington, D.C.: Urban Institute Press.

Borkowski, Nancy. 2005. "Organizational Behavior in Health Care." In Content Theories of Motivation, edited by Nancy Borkowski, pp. 113-138. Boston: Jones & Bartlett.

Bradner, Jeanne H. 1997. "Volunteer Management." In The Nonprofit Handbook: Management, edited by Tracy Daniel Connors, pp. 162-192. New York: John Wiley.

Brody, Evelyn. 2002. "Accountability and Public Trust." In The State of Nonprofit America, edited by Lester M. Salamon, pp. 471-498. Washington, D.C.: Brookings Institution Press.

Brudney, Jeffrey L. 1999. "The Effective Use of Volunteers: Best Practices for the Public Sector." Law & Contemporary Problems 62 (autumn):219-254.

Bryce, Herrington J. 2005. Players in the Public Policy Process: Nonprofits as Social Capital and Agents. New York: Palgrave Macmillan.

Causer, Craig. 2004. "Special Report: Nonprofits Building Trust Through Transparency." NonProfit Times, November 1 (www.nptimes.com/Nov04/sr4.html).

Charity Lobbying in the Public Interest. n.d. "The Basics: Nonprofit Advocacy and Lobbying" (www.clpi.org/nonprofitadvocacyandlobbying_Basics.aspy).

Charity Navigator. n.d. "10 of the Best Charities Everyone's Heard Of" (www.charitynavigator.org/index.cfm/bay/topten.detail/listid/18.htm).

Coffee, Peter. 2002. "The Hierarchy of Needs in High Tech." eWeek 19, no. 27:43.

Conservation International (CI). 2006 "CI's Mission." http://web.conservation.org/ xp/CIWEB/about/).

Cullen, Dallas, and Lise Gotell. 2002. "From Orgasms to Organizations: Maslow, Women's Sexuality and the Gendered Foundations of the Needs Hierarchy." Gender, Work and Organization 9, no. 5:537-555.

Davis, Andrea Muirragui. 2004. "NFP Loans to Insiders Raise Legal Questions: Indiana One of 19 States to Prohibit or Limit Practice." Indianapolis Business Journal 24.1.

Davis, James H., F. David Schoorman, and Lex Donaldson. 1997. "Toward a Stewardship Theory of Management." Academy of Management Review 22, no.

1:20-47.

Duffy, Karen G., and James Thomson. 1992. "Community Mediation Centers: Humanistic Alternatives to the Court System." Journal of Humanistic Psychology 32 (spring):101-114.

Franklin, Robert. 2004. "Some Nonprofits Make Loans Close to Home; Executives, Board Members Are Recipients." Minneapolis Star Tribune, February 14:1A.

Frumkin, Peter. 2002. On Being Nonprofit: A Conceptual and Policy Primer. Cambridge, Mass.: Harvard University Press.

Gormley, Michael. 2003. "Spitzer: Charity Telemarketers Pocket $2 For Every $3 Raised." Associated Press State & Local Wire [Albany, N.Y.], December 27 (http://0-web.lexis-nexis.com.wncln.wncln.org/universe).

———. 2004. "Telemarketers Take Up to 90 Percent Cut in Charitable Donations." Associated Press State & Local Wire [Albany, N.Y.], December 24 (http://0-web.lexis-nexis.com.wncln.wncln.org/universe).

Grace, Kay Sprinkel. 2006. Beyond Fundraising: New Strategies for Nonprofit Innovation and Investment. 2nd ed. Hoboken, N.J.: John Wiley.

Halepota, Hassan Ali. 2005. "Motivational Theories and Their Application in Construction." Cost Engineering 47 (March):14-18.

Hananel, Sam. 2004. "Charities Urged to Change to Restore Trust." PhillyBurbs.com (November 22).

Hansmann, Henry B. 1980. "The Role of Nonprofit Enterprise." Yale Law Journal 89 (April 1980): 835-98.

Haruna, Peter Fuseini. 2000. "An Empirical Evaluation of Motivation and Leadership Among Career Public Administrators: The Case of Ghana." Ph.D. dissertation, University of Akron.

Hatwick, Richard E. 1986. "The Behavioral Economics of Business Ethics." Journal of Behavioral Economics 15 (spring/summer):87-101.

Hendershott, Anne, Patrick Drinan, and Megan Cross. 2000. "Toward Enhancing a Culture of Academic Integrity." NASPPA Journal 37, no. 4:587-597.

IRS (Internal Revenue Service). n.d. " Life Cycle of a Public Charity" (www.irs.gov/charities/article/0,,id=12260,00.html). "Charities and Non-Profits Exemption Requirements" (www.irs.gov/charities/charitable/article/0,,id=96099,00.html).

———. 2006. "Life Cycle of a Public Charity" (www.irs.gov/charities/article/0,,id=12260,00.html).

Ivashchenko, A., and D. Novikov. 2006. "Model of the Hierarchy of Needs." Automation and Remote Control 67, no. 9:1512-1517.

Jackson, Peggy M., and Toni E. Fogarty. 2005. Sarbanes-Oxley for Nonprofits: A Guide to Gaining the Competitive Advantage. Hoboken, N.J.: John Wiley.

Janiszewski, Randee. 2005. "Motivational Factors That Influence Baby Boomers Versus Generational X: Independent Insurance Agents." Ph.D. dissertation, Capella University.

Jeavons, Thomas H. 2005. "Ethical Nonprofit Management." In The Jossey-Bass Handbook of Nonprofit Leadership and Management, 2nd ed., edited by Robert D. Herman and associates, pp. 204-229. San Francisco: John Wiley.

Johnson, Chip. 2004. "Watching the Police's Watchdogs." San Francisco Chronicle, August 13:B1.

Latham, Gary P., and Craig C. Pinder. 2005. "Work Motivation Theory and Research at the Dawn of the Twenty-first Century." Annual Review of Psychology 56:485-516.

Lord, Robert L. 2002. "Traditional Motivation Theories and Older Engineers." Engineering Management Journal 14 (September):3-8.

Macduff, Nancy. 2005. "Principles of Training for Volunteers and Employees." In The Jossey-Bass Handbook of Nonprofit Leadership and Management, 2nd ed., edited by Robert D. Herman and associates, pp. 703-730. San Francisco: John Wiley.

Maslow, Abraham H. 1943. "A Theory of Human Motivation." Psychological Review 50, no. 4:370-396.

McCabe, D. L., and L. K. Trevino. 1996. "What We Know About Cheating in College." Change 28, no. 1:28-34.

McCurley, Stephen. 2005. "Keeping the Community Involved: Recruiting and Retaining Volunteers." In The Jossey-Bass Handbook of Nonprofit Leadership and Management, 2nd ed., edited by Robert D. Herman and associates, pp. 587-622. San Francisco: John Wiley.

NCF (National Christian Foundation). 2006. "Smart Christian Giving" (www.nationalchristian.com).

Osborn, Claire. 2004. "Two Must Pay Charity $14 Million; Damages Are to Punish Former King Foundation Executives for Bloated Salaries, Credit Card Use." Austin American-Statesman, June 15:B1.

Ostrower, Francie, and Marla J. Bobowick. 2006. "Nonprofit Governance and the Sarbanes-Oxley Act" (www.boardsource.org/dl.asp?document_id=473).

Owen, John R., III. 2003. "Keeping Nonprofits Clean; An Audit Committee Is Needed to Help the Board Do Its Job Correctly." Pittsburgh Post-Gazette, April 8:D2.

Pisello, Tom. 2003. "Anticipating IT Needs in Pyramidal Steps." Computerworld 37, no. 34:49.

Rouse, Kimberly A. Gordon. 2004. "Beyond Maslow's Hierarchy of Needs: What Do People Strive For?" Performance Improvement 43, no. 10:27-31.

Salamon, Lester. 1995. Partners in Public Service. Baltimore: Johns Hopkins University Press.

———. 2005. Johns Hopkins Nonprofit Listening Post Project: Nonprofit Financial Disclosure. Communiqué No. 4 (www.jhu.edu/listeningpost/news/pdf/comm04.pdf).

Salmon, Jacqueline L. 2004. "Nonprofit Endorsements Will Expand: Maryland Group to Certify Charities Across Nation." Washington Post, June 27:C01.

Saltonstall, David, and Heidi Evans. 2004. "Hale House Chief Quits After 22 Months." New York Daily News, March 31:32.

Schmidt, Elizabeth. 2004. "How Ethical Is Your Nonprofit Organization?" (www.guidestar.org/news/features/ethics.jsp).

Schwing, Richard. 2002. "A Mental Model Proposed to Address the Sustainability and Terrorism Issues." Risk Analysis 22, no. 3:415-420.

"Shelter's Ex-Director Sentenced for Stealing." 2002. Milwaukee Journal Sentinel October 25:09A.

Sinclair, Matthew. 2004. "Nonprofit Whistleblowers Need Protection." NonProfit Times (wwwnptimes.com/Jun04/npt1.html).

Smucker, Bob. 1999. The Nonprofit Lobbying Guide. 2nd ed. Washington D.C.: Independent Sector.

Soltz, Barbara A. Burgess. 1997. "The Board of Directors." In The Nonprofit Handbook: Management, 2nd ed., edited by Tracy Daniel Connors, pp. 114-147. New York: John Wiley.

Souccar, Miriam Kreinan. 2004. "Nonprofit Hails New Director for Hardy Children's Charity; Will Help Hale House Rebuild After Scandal?" Crain's New York Business, April 12:14.

Stephens, Joe. 2004. "Overhaul of Nature Conservancy Urged. Report by Independent Panel Calls for Greater Openness." Washington Post, March 31:A01.

Strom, Stephanie. 2003. "Accountability; New Equation for Charities: More Money, Less Oversight." New York Times, November 17:F1.

"Theft from Cancer Society Raises Several Questions." 2000. Columbus Dispatch, July 25:8A.

Tyler, J. Larry, and Errol L. Biggs. 2004. "Conflict of Interest: Strategies for Remaining 'Purer Than Caesar's Wife.' " Trustee 57 (March):22-26.

Vigil, Jennifer. 2004. "$3.9 Million Spent Directly for Victims." San Diego Union-Tribune, May 2 (http://signonsandiego.com/news/fires/20040502-9999-/lm2redfunds.html/).

Watson, Charles T. 2000. "Accountability a Key Factor for Groups Soliciting Help." Ventura County Star, July 16:E06.

Watson, Mary R., and Rikki Abzug. 2005. "Finding the Ones You Want, Keeping the Ones You Find: Recruitment and Retention in Nonprofit Organizations." In The Jossey-Bass Handbook of Nonprofit Leadership and Management, 2nd ed., edited by Robert D. Herman and associates, pp. 623-659. San Francisco: John Wiley.

Weiner, Stanley. 2003. "Proposed Legislation: Its Impact on Not-For-Profit Board Governance." CPA Journal 73 (November):56 ff. (http://web.lexis-nexis.com/universe/).

Werther, William B., Jr., and Evan M. Berman. 2001. Third Sector Management: The Art of Managing Nonprofit Organizations. Washington, D.C.: Georgetown University Press.

Wymer, Walter, Jr., Patricia Knowles, and Roger Gomes. 2006. Nonprofit Marketing: Marketing Management for Charitable and Nongovernmental Organizations. Thousand Oaks, Calif.: Sage.

Young, Dennis R., and Lester Salamon. 2002. "Commercialization, Social Ventures, and For-Profit Competition." In The State of Nonprofit America, edited by Lester M. Salamon, pp. 423-446. Washington, D.C.: Brookings Institution Press.

ABOUT THE AUTHORS

Ruth Ann Strickland is a professor in the Political Science and Criminal Justice Department at Appalachian State University, teaching and conducting research on public personnel administration, ethics in nonprofit management, and public policy analysis. She obtained her Ph.D. in political science in 1989 at the University of South Carolina. She has published four books, nine book chapters, and more than twenty peer-reviewed articles in a wide array of journals.

Shannon K. Vaughan is an assistant professor and assistant M.P.A. director in the Political Science and Criminal Justice Department at Appalachian State University, teaching and conducting research on not-for-profit organizations, ethics, and policy analysis. She obtained her Ph.D. in political science in 2004 at the University of Kentucky. Her field experience includes positions as a nonprofit executive director and as a grants specialist, where she prepared grant proposals that generated more than $4.7 million for government and non-profit organizations.

From: Ruth Ann Strickland and Shannon K. Vaughan, "The Hierarchy of Ethical Values in Nonprofit Organizations: A Framework for an Ethical, Self-Actualized Organization Culture," *Public Integrity* 10, no. 3 (2008): 233-251. Used with permission.

13.4 The New Look of Transparency

By Kristin Clarke

OBJECTIVES:

15. Define the term "transparent" as it refers to organizations.
16. Name the finance report that nonprofit agencies (including CAP) are required to publish annually.
17. Identify benefits of having transparent communication within an organization.
18. List steps that an organization can take to become more transparent.

A board member calls for a meeting to move into executive session. Under what circumstances do you, as CEO, voice opposition?

The membership department receives several inquiries about the percentage gap between your CEO's compensation and that of your lowest-paid employee. Do you share that information?

A potential donor asks your organization to provide a copy of its whistleblower policy. Do you have one?

These examples are real. Members, donors, media, regulators, the public, and volunteers are just some of the stakeholders whose demand for greater transparency and its close cousin, accountability, has grown in the past decade.

Finances, of course, top the list for scrutiny, followed closely by governance and communication. The corporate world has been coping with a new era of regulated transparency and accountability ever since passage of the Sarbanes-Oxley Act (SOX) in 2002. Some nonprofits and associations, concerned that the law would expand to their sector as well, directed their auditors, investment committees, and boards to voluntarily adopt similar governance and accounting principles.

According to SOX coauthor and former senator Mike Oxley, airing the inner workings of nonprofits was never part of any discussions by lawmakers. He applauds such initiative, though, and urges other associations to follow suit. Speaking at the 2010 Council for Non-Profit Accountability Summit, Oxley stated that such activities "will improve the fiscal condition of nonprofits and strengthen donor confidence."

And Congress may yet change its mind about the scope of SOX. In a later interview, Oxley warned, "A bit of caution on the part of the nonprofits and some planning hopefully will mean that down the road they won't have to face this kind of problem, because once you have that breach of reputation risk, boy, it can go downhill very, very fast."

Says Ron Noden, chair of the Council for Non-Profit Accountability, "[Transparency] is an issue that will continue to get attention in the nation's capital and in state houses around the country. We need to be proactive, so nonprofits can continue to be mission focused."

WHAT DOES TRANSPARENCY LOOK LIKE?

A major hurdle, though, is the cloudy definition of a "transparent organization." Warren Bennis, founder of The Leadership Institute at the University of Southern California, wrote an entire book on the subject, Transparency: How Leaders Create a Culture of Candor, and still acknowledges that the term "has many different meanings" and has evolved in the past 10 years.

"One of those meanings is the transparency of transactions ... [N]ot having enough of that led to the recent [financial] crash," Bennis says, adding, "The word 'transparency' in the business lexicon and in the vernacular I'm familiar with has everything to do with how open, how visible, organizations are in dealing with various stakeholders and also within the organization—how transparent our people are with each other, how candid they are."

Bennis and his coauthors emphasize that the burden and opportunities around greater transparency are here to stay because of the multiple information outlets now available to consumers, especially online.

"We can find out who the best practitioners are in almost any particular branch of medicine or profession [just by visiting a few websites], so to the extent that people are educated and can distinguish new sources, it's going to be very hard to keep things secret unless there is some kind of federal provision or patent law that would [do so]," he says.

Bennis recalls a 2005 speech he gave at Harvard University called "Transparency Is Inevitable." At the time, he estimates, only 20 percent of the audience had ever heard of the word "blogosphere." Now, thousands of blogs, not to mention microblogs via Twitter, are born daily.

"Almost every company is going to be under the gun about the problems of not being transparent enough," Bennis warns. He adds, "Look at what happened with Toyota by trying to keep [safety issues] quiet, or Merck [whose antiarthritis medication Vioxx was pulled over safety concerns]—billions of dollars of penalties and losses of customer support."

Because of the high stakes, Bennis urges leaders to work harder to better understand the issue and ask tough questions. "They need to know about the whole revolution in social networking and networking media because of what is going on there," he says. "That's the key thing. It behooves organizations to be as transparent as [possible] without giving away trade secrets."

What if they aren't comfortable lifting the cloak? What if they don't even see the cloak? "Just look at Enron," Bennis says. "Look at any recent story on whistleblowers. The risks are enormous and are increasing every day given the number of sources we have and the changing nature of how we get information right now. The ... risks of not having some kind of transparency policy are very—well, I don't think it's worth it."

THE COMMUNICATION CONUNDRUM

One of the highest-profile moves toward greater openness in the association and nonprofit world has been the recently revised IRS Form 990. Calling the updated form "a major step in transparency," charity tracker GuideStar cautioned association leaders in June 2009: "The impact that the increased transparency will have on nonprofit organizations has been severely underestimated. It is not sufficient for nonprofit staff and board members simply to be made aware of these changes. They must also be alert to the changes' strategic implications and have tools to manage them successfully."

That requires good governance, agree GuideStar and others, including public clarity about how board nominations occur, are vetted, and are executed; how the board and CEO make decisions; how money is allocated; and how the mission is progressing.

Association finance committees appear to be drawing special scrutiny. Who are these people? How were they chosen? How do they make decisions about association investments? One association professional recalls serving on a board that refused to even second her motion to discuss, much less act on, moving investments from companies with major Clean Water Act violations—even though the organization's mission includes clean water advocacy. Those companies were providing good returns, the board responded. Would most members have agreed to set aside mission in favor of profit?

Some additional concerns of transparency proponents are weak communication access, content, and delivery, as well as perceptions around stakeholder inclusiveness. Associations are now experimenting with new ways to meet member transparency expectations, whether by adopting virtual tools for collaborative note taking and all-access post-meeting discussions, tweeting live from events, or uploading recorded meetings to free or paid-access archives.

Jeffrey Solomon, executive director of Andrea and Charles Bronfman Philanthropies Inc. and author of the book *The Art of Giving*, even suggests live streaming your board meetings on the internet.

"Why not?" he asks.

Maybe because of the sensitivity of some issues up for debate or worries about directors posturing for cameras? When several nonprofit CEOs heard that suggestion, reactions ranged from snorts to sighs to grimaces. "That could be ugly, but I do wonder if it would help keep everyone more focused on the job at hand," says one longtime leader, who asked to remain anonymous out of concern for how his comments might be perceived by his board.

Less ticklish are engagement tactics such as adding reader ratings to online articles a la Amazon or reorganizing web content for easier access.

But public evaluations of association speakers, education sessions, or even attendees' overall conference experiences? That could cause some squirms. What about website usage stats such as those provided by the "Green" Hotels Association, which wanted members to see the growth in visitorship to its site? Would an organization take those stats down if the numbers start dropping?

And considering how little time members claim they have, when does it all become too much information anyway? There are costs involved in sharing, complain leaders. Staff time, for instance, or the expense of building new web systems or sites.

But there the benefits of transparency can also add up. In his book, *Straight A Leadership: Alignment, Action, Accountability*, healthcare leader Quint Studer discusses the vital role of transparency in creating a successful workplace culture.

"Leaders have talked about transparency for a long time, but it's never been more important than it is now," says Studer. "Remember, we share information with employees for a couple of reasons: One, it's the right thing to do, and two, it's good for business. And most companies can use every possible edge these days."

He cites the benefits of a work culture of free-flowing information: a greater connection by staff to the financial big picture, reduced complacency, more creative solutions, and "organizational consistency and stability and faster, more-efficient execution." All of that helps organizations compete, especially in a weak economy, he says.

Bennis agrees that a workplace that recognizes the sound business case for transparency is essential for leaders to surmount the challenges of crafting a relevant strategy. "In the long run it would be an enormous advantage for an organization," he says. "The difficulties are that [a transparency policy] would have to be adjusted to each organization [because it] has enormous implications for their ethics and values, and how those are enforced. ... Given the fact that inside of organizations are things going on that the public should know about, [stakeholders] are not shutting up."

Associations Open Up

Some associations have looked to transparency as a way to push their mission, build donor trust, boost engagement and dialogue with members, address regulator concerns, and modernize their risk-management strategies.

- The Washington State Hospital Association and its member hospitals launched a webpage called "Hospital Transparency" to help consumers make healthcare decisions, learn about costs and quality measurements of hospital care, and identify nearby facilities and financial assistance options.

- The Oregon Association of Hospitals and Health Systems partnered with the Office of Health Policy and Research to release a report in May 2010 that makes public the hospital-acquired infection rates of health facilities in the state. According to Steve Gordon, Ph.D., of the association's quality committee in The Lund Report, "the intent is to be transparent" and "to use [the report] as a foundation for continued prevention."

- The National Association of Corporate Directors used transparency to promote the value of its programs, publicly reaffirming the importance of and its commitment to director education: "At a time when new SEC disclosure rules call for greater transparency of board member qualifications ... [we] will continue to provide the industry's leading certificate-based director education and in-boardroom services for the largest and most complex companies around the world, as well as all publicly traded, private, and nonprofit companies."

- ASCD (formerly the Association for Supervision and Curriculum Development) has turned to the virtual platform Skype to support more inclusive, open meetings of its Scholars Team, whose 25 members reside on six continents. The free tool can record meetings, so ASCD can offer them archived online later.

- International relief nonprofit World Vision and others issued frequent updates to donors and media about the exact uses and on-the-ground impacts of the millions of dollars donated after the Haiti earthquake in January 2010.

SEVEN STEPS TO A MORE TRANSPARENT ORGANIZATION

Here's how you can create a more transparent organization:

1. Make sure senior leadership is aligned. Does everyone see the external environment the same way? Does everyone understand organizational goals and plans? Does everyone agree on what success looks like? If not, it's time to remedy the situation.

"Alignment is most important at the senior level because all information cascades downward from it," says Studer. "If one senior leader is out of sync with the others, then everyone under her is going to be out of sync."

2. Close the perception gap between senior leadership and middle managers. Senior leaders generally have a clear grasp of the issues facing the organization. They are steeped in these issues every day. Mid-level managers don't always see things the same way. The only solution is for senior leaders to relentlessly communicate the issues to them.

"You can address these issues in supervisory sessions," suggests Studer. "You can hold regular meetings with mid-level managers. You can send out email alerts that link to news items driving high-level decisions. If you're a senior leader, it's critical to make sure the people under you understand the big-picture issues and their implications. It's one of the most important parts of your job."

3. Help people understand the true financial impact of decisions. Get comfortable framing all major decisions in economic terms. If a manager wants to spend money on something—a new program, a new position—she needs to be prepared to explain in financial terms how it will pay off for the company. Staff, too, need to understand the real cost of mistakes or lapses in productivity as well as the potential positive impact of doing things in a new way.

"Many of the healthcare leaders I work with use a financial impact grid to educate employees on how certain issues translate to dollars," says Studer. "The idea is to teach everyone to think like the CFO. Educating people in this way can be very powerful in changing their behavior."

4. Put mechanisms in place for communicating vital issues to frontline employees. People aren't going to pick up on what leaders want them to know by osmosis. You need to tell them clearly, succinctly, and often. That means putting in place a system, or a series of systems, to ensure that transparency gets translated into action.

5. Prepare managers to answer tough questions. If managers tell staff the organization is instituting a hiring or salary freeze, they'll almost certainly hear questions like, "If money's so tight, how can the company afford the new database?" The manager needs to know ahead of time exactly how to answer, so he won't blurt out a we/they perpetuator like, "Sorry, that's the orders from the top."

"In a transparent [organization], there's no reason to hide financial realities from anyone—but that doesn't mean managers naturally know the best way to phrase their answers," says Studer. "Some are just better communicators than others. Anticipating tough questions, formulating the right key words, and sharing them with leaders at all levels allows everyone to answer them consistently."

6. When you have bad news, treat employees like adults. Once a tough decision has been made, share it with everyone immediately. "Knowing what's happening and what it means is always better than not knowing," says Studer. "And often, what people are imagining is worse than what's really happening."

7. Keep people posted. When something changes, let employees know. This builds trust between leaders and staff and keeps them connected to the big picture.

"Be sure to share any good news you get," says Studer. "Transparency doesn't mean all bad news, all the time. When you disseminate positive developments as quickly as you do negative ones, you boost employee morale and reinforce any progress that's being made."

ABOUT THE AUTHOR

Kristin Clarke is a writer and researcher for ASAE. Email: kclarke@asaecenter.org

From: Kristin Clarke, "Clear: The New Look of Transparency," Associations Now (November 2010), used with permission.

13.5 Public and Private Management:

Are They Fundamentally Alike in All *Unimportant* Respects?

By Graham T. Allison

OBJECTIVES:

19. Define the term "management" as used in this article.
20. Name the organizations and managers this author uses to contrast the differences in public and private management.
21. List the functions of general management, as identified by Allison.
22. Identify key differences in managing public and private organizations.
23. Describe the constitutional difference between public and private management.

My subtitle puts Wallace Sayre's oft quoted "law" as a question. Sayre had spent some years in Ithaca helping plan Cornell's new School of Business and Public Administration. He left for Columbia with this aphorism: public and private management are fundamentally alike in all unimportant respects.

Sayre based his conclusion on years of personal observation of governments, a keen ear for what his colleagues at Cornell (and earlier at OPA) said about business, and a careful review of the literature and data comparing public and private management. Of the latter there was virtually none. Hence, Sayre's provocative "law" was actually an open invitation to research.

Unfortunately, in the 50 years since Sayre's pronouncement, the data base for systematic comparison of public and private management has improved little...I, in effect, like to take up Sayre's invitation to speculate about similarities and difference among public and private management in ways that suggest significant opportunities for systematic investigation...

FRAMING THE ISSUE: WHAT IS PUBLIC MANAGEMENT?

What is the meaning of the term *management* as it appears in *Office of Management and Budget* or *Office of Personnel Management*? Is "management" different from, broader, or narrower than "administration"? Should we distinguish between management, leadership, entrepreneurship, administration, policy making, and implementation?

Who are "public managers"? Mayors, governors, and presidents? City managers, secretaries, and commissioners? Bureau chiefs? Office directors? Legislators? Judges?

Recent studies of OPM and OMB shed some light on these questions. OPM's major study of the "current status of public management research" completed in May 1978 by Selma Mushkin and colleagues of Georgetown's Public Service Laboratory starts with this question. The Mushkin report notes the definition of *public administration* employed by the Interagency Study Committee on Policy Management Assistance in its 1975 report to OMB. That study identified the following core elements:

1. *Policy Management*. The identification of needs, analysis of options, selection of programs, and allocation of resources on a jurisdiction-wide basis.

2. *Resource Management*. The establishment of basic administrative support systems, such as budgeting, financial management, procurement and supply, and personnel management.

3. *Program Management*. The implementation of policy of daily operation of agencies carrying out policy along functional lines (education, law enforcement, etc.).[1]

The Mushkin report rejects this definition in favor of an "alternative list of public management elements." These elements are:

- Personnel management (other than work force planning, collective bargaining and labor relations)
- Work force planning
- Collective bargaining and labor-management relations
- Productivity and performance measurement
- Organization/reorganization
- Financial management (including the management of intergovernmental relations)
- Evaluation research, and program and management audit.[2]

Such terminological tangles seriously hamper the development of public management as a field of knowledge.

In our efforts to discuss the public management curriculum at Harvard, I have been struck by how differently people use these terms, how strongly many individuals feel about some distinction they believe is marked by a difference between one word and another, and consequently, how large a barrier terminology is to convergent discussion. These verbal obstacles virtually prohibit conversation that is both brief and constructive among individuals who have not developed a common language or a mutual understanding of each other's use of terms.

This terminological thicket reflects a more fundamental conceptual confusion. There exists no overarching framework that orders the domain. In an effort to get a grip on the phenomena – the buzzing, blooming confusion of people in jobs performing tasks that produce results – both practitioners and observers have strained to find distinctions that facilitate their work. The attempts in the early decades of this century to draw up a sharp line between "policy" and "administration," like more recent efforts to mark a similar divide between "policy-making" and "implementation," reflect a common search for a simplification that allows one to put the value-laden issues of politics to one side (who gets what, when, and how), and focus on the more limited issue of how to perform tasks more efficiently.[3] But can anyone really deny that the "how" substantially affects the "who," the "what," and the "when"? The basic categories now prevalent in discussion of public management – strategy, personnel management, financial management, and control – are mostly derived from a business context in which executives manage hierarchies. The fit of these concepts to the problems that confront public managers is not clear.

Finally, there exist no ready data on what public managers do. Instead, the academic literature, such as it is, mostly consists of speculation tied to bits and pieces of evidence about the tail or the trunk or other manifestation of the proverbial elephant.[4] In contrast to the literally thousands of cases describing problems faced by private managers and their practice in solving these problems, case research from the perspective of a public manager is just beginning...[5] The paucity of data on the phenomena inhibits systematic empirical research on similarities and differences between public and private management, leaving the field to a mixture of reflection on personal experience and speculation.

For the purpose of this presentation, I will follow Webster and use the term *management* to mean the organization and direction of resources to achieve a desired result. I will focus on *general managers*, that is, individuals charged with managing a whole organization or multi-functional subunit. I will be interested in the general manager's full responsibilities, both *inside* his organization in integrating the diverse contributions of specialized subunits of the organization to achieve results, and *outside* his organization in relating his organization and its product to external constituencies. I will begin with the simplifying assumption that managers of traditional government organizations are public managers, and managers of traditional private businesses [are] private managers. Lest the discussion fall victim to the fallacy of misplaced abstraction, I will take the Director of EPA and the Chief Executive Officer of American Motors as, respectively, public and private managers. Thus, our central question can be put concretely: in what ways are the jobs and responsibilities of Doug Costle as Director of EPA similar to and different from those of Roy Chapin as Chief Executive Officer of American Motors?

SIMILARITIES: HOW ARE PUBLIC & PRIVATE MANAGEMENT ALIKE?

At one level of abstraction, it is possible to identify a set of general management functions. The most famous such list appeared in Gulick and Urwick's classic *Papers in the Science of Administration*.[6] [They] summarized the work of the chief executives in the acronym POSDCORB. The letters stand for:

- Planning
- Organizing
- Staffing
- Directing
- Coordinating
- Reporting
- Budgeting

With various additions, amendments, and refinements, similar lists of general management functions can be found through the management literature from Barnard to Drucker.[7]

I shall resist here my natural academic instinct to join the intramural debate among proponents of various lists and distinctions. Instead, I simply offer one composite list (see Table 1) that attempts to incorporate the major functions that have been identified for general managers, whether public or private.

These common functions of management are not isolated and discrete, but rather integral components separated here for purposes of analysis. The character and relative significance of the various functions differ from one time to another in the history of any organization, and between one organization and another. But whether in a public or private setting, the challenge for the general manager is to integrate all these elements so as to achieve results.

TABLE 1: FUNCTIONS OF GENERAL MANAGEMENT

Strategy

1. *Establishing objectives and priorities* for the organization (on the basis of forecasts of the external environment and the organization's capacities).

2. *Devising operational plans* to achieve these objectives.

Managing Internal Components

3. *Organizing and staffing.* In organizing the manager establishes structure (units and positions with assigned authority and responsibilities) and procedures for coordinating activity and taking action. In staffing he tries to fit the right persons in the key jobs.*

4. *Directing personnel and the personnel management system.* The capacity of the organization is embodied primarily in its members and their skills and knowledge. The personnel management system recruits, selects, socializes, trains, rewards, punishes, and exits the organization's human capital, which constitutes the organization's capacity to act to achieve its goals and to respond to specific directions from management.

5. *Controlling performance.* Various management information systems – including operating and capital budgets, accounts, reports, and statistical systems, performance appraisals, and product evaluation – assist management in making decisions and in measuring progress towards objectives.

Managing External Components

6. *Dealing with "external" units* of the organization subject to some common authority. Most general managers must deal with general managers of other units within the larger organization – above, laterally, and below – to achieve their unit's objectives.

7. *Dealing with independent organizations.* Agencies from other branches or levels of government, interest groups, and private enterprises that can importantly affect the organization's ability to achieve its objectives.

8. *Dealing with the press and the public* whose action or approval or acquiescence is required.

*Organization and staffing are frequently separated in such lists, but because of this interaction between the two, they are combined here. See Graham Allison and Peter Szanton, *Remaking Foreign Policy* (New York: Basic Books, 1976), p. 14.

DIFFERENCES: HOW ARE PUBLIC & PRIVATE MANAGEMENT DIFFERENT?

While there is a level of generality at which management is management, whether public or private, functions that bear identical labels take on rather different meanings in public and private settings. As Larry Lynn has pointed out, one powerful piece of evidence in the debate between those who emphasize "similarities" and those who underline "differences" is the nearly unanimous conclusion of individuals who have been general managers in both business and government. Consider the reflections of George Shultz (Secretary of State; former Director of OMB, Secretary of Labor, Secretary of the Treasury, President of Bechtel), Donald Rumsfeld (former congressman, Director of OEO, Director of the Cost of Living Council, White House Chief of Staff, and Secretary of Defense; now President of G. D. Searle and Company), Michael Blumenthal (former Chairman and Chief Executive Officer of Bendix, Secretary of the Treasury, and now Vice Chairman of Burroughs), Roy Ash (former President of Litton Industries, Director of OMB; later President of Addressograph), Lyman Hamilton (former Budget Officer in BOB, High Commissioner of Okinawa, Division Chief in the World Bank and President of ITT), and George Romney (former President of American Motors, Governor of Michigan, and Secretary of Housing and Urban Development.)[8] All judge public management different from private management – and harder!

Orthogonal Lists of Differences

My review of these recollections, as well as the thoughts or academics, has identified three interesting, orthogonal lists that summarize the current state of the field: one by John Dunlop; one major *Public Administration Review* survey of the literature comparing public and private organizations by Hal Rainey, Robert Backoff and Charles Levine; and one by Richard E. Neustadt, prepared for the National Academy of Public Administration's Panel on Presidential Management.

John T. Dunlop's "impressionistic comparison of government management and private business" yields the following contrasts.[9]

1. ***Time Perspective.*** Government managers tend to have relatively short time horizons dictated by political necessities and the political calendar, while private managers appear to take a longer time perspective oriented toward market developments, technological innovation and investment, and organization building.

2. ***Duration.*** The length of service of politically appointed top government managers is relatively short, averaging no more than 18 months recently for assistant

secretaries, while private managers have a longer tenure both in the same position and in the same enterprise. A recognized element of private business management is the responsibility to train a successor or several possible candidates, [whereas] the concept is largely alien to public management, since fostering a successor is perceived to be dangerous.

3. ***Measurement of Performance.*** There is little if any agreement on the standards and measurement of performance to appraise a government manager, while various tests of performance – financial return, market share, performance measures for executive compensation – are well established in private business and often made explicit for a particular managerial position during a specific period ahead.

4. ***Personnel Constraints.*** In government there are two layers of managerial officials that are at times hostile to one another: the civil service (or now the executive system) and the political appointees. Unionization of government employees exists among relatively high-level personnel in the hierarchy and includes a number of supervisory personnel. Civil service, union contract provisions, and other regulations complicate the recruitment, hiring, transfer, and layoff or discharge of personnel to achieve managerial objectives or preferences. By comparison, private business managements have considerably greater latitude, even under collective bargaining, in the management of subordinates. They have much more authority to direct the employees of their organization. Government personnel policy and administration are more under the control of staff (including civil service staff outside an agency) compared to the private sector in which personnel are much more subject to line responsibility.

5. ***Equity and Efficiency.*** In governmental management great emphasis tends to be placed on providing equity among different constituencies, while in private business management relatively greater stress is placed upon efficiency and competitive performance.

6. ***Public Processes versus Private Processes.*** Governmental management tends to be exposed to public scrutiny and to be more open, while private business management is more private and its processes more internal and less exposed to public review.

7. ***Role of Press and Media.*** Governmental management must contend regularly with the press and media; its decisions are often anticipated by the press. Private decisions are less often reported in the press, and the press has a much smaller impact on the substance and timing of decisions.

8. ***Persuasion and Direction.*** In government, managers often seek to mediate decisions in response to a wide variety of pressures and must often put together a coalition of inside and outside groups to survive. By contrast, private management process much more by direction or the issuance of orders to subordinates by superior managers with little risk of contradiction. Governmental managers tend to regard themselves as responsive to many superiors, while private managers look more to one higher authority.

9. ***Legislative and Judicial Impact.*** Governmental managers are often subject to close scrutiny by legislative oversight groups or even judicial orders in ways that are quite uncommon in private business management. Such scrutiny often materially constrains executive and administrative freedom to act.

10. ***Bottom Line.*** Governmental managers rarely have a clear bottom line, while that of a private business manager is profit, market performance, and survival.

The *Public Administration Review's* major review article comparing public and private organizations, [by Rainey, Backoff and Levine,] attempts to summarize the major points of consensus in the literature on similarities and differences among public and private organizations.[10]

Third, Richard E. Neustadt, in a fashion close to Dunlop's, notes six major differences between Presidents of the United States and Chief Executive Officers of major corporations.[11]

1. ***Time Horizon.*** The private chief begins by looking forward a decade, or thereabouts, his likely span barring extraordinary troubles. The first term president looks forward four years at most, with the fourth (and now even the third) year dominated by campaigning for reelection (what second-termers look toward we scarcely know, having seen but one such term completed in the past quarter century).

2. ***Authority over the Enterprise.*** Subject to concurrence from the Board of Directors which appointed and can fire him, the private executive sets organization goals, shifts structures, procedures, and personnel to suit, monitors results, reviews key operations decisions, deals with key outsiders, and brings along his Board. Save for the deep but narrow sphere of military movements, a president's authority in these respects is shared with well-placed members of Congress (or their staffs): case by case, they may have more explicit authority that he does (contrast authorizations and appropriations with the "take-care" clause). As for "bringing along the Board," neither the congressmen with whom he shares power nor the primary and general electorates which "hired" him have either a Board's duties or a broad view of the enterprise precisely matching his.

3. Career System. The model corporation is a true career system, something like the Forest Service after initial entry. In normal times the chief himself is chosen from within, or he is chosen from another firm in the same industry. He draws department heads (and other key employees) from among those with whom he's worked or whom he knows in comparable companies. He and his principal associates will be familiar with each other's roles – indeed, he probably has had a number of them – and also usually with one another's operating styles, personalities, idiosyncrasies. Contrast the president who rarely has had much experience "downtown," probably knows little of most roles there (much of what he knows will turn out wrong), and less of most associates whom he appoints there, willy nilly, to fill places by Inauguration Day. Nor are they likely to know one another well, coming as they from "everywhere" and headed as most are toward oblivion.

4. Media Relations. The private executive represents his firm and speaks for it publicly in exceptional circumstances; he and his associates judge the exceptions. Those aside, he neither sees the press nor gives its members access to internal operations, least of all in his own office, save to make a point deliberately for public-relations purposes. The president, by contrast, is routinely on display, continuously dealing with the White House press and with the wider circle of political reporters, commentators, columnists. He needs them in his business, day by day, mothering exceptional about it, and they need him in theirs: the TV network news programs lead off with him some nights each week. They and the president are as mutually dependent as he and congressmen (or more so). Comparatively speaking, these relations overshadow most administrative ones much of the time for him.

5. Performance Measurement. The private executive expects to be judged, and in turn to judge subordinates, by profitability, however the firm measures it (a major strategic choice). In practice, his Board may use more subjective measures; so may he, but at risk to morale and good order. The relative virtue of profit, of "the bottom line," is its legitimacy, its general acceptance in the business world by all concerned. Never mind its technical utility in given cases; its apparent "objectivity," hence "fairness," has enormous social usefulness: a myth that all can live by. For a president there is no counterpart (except, *in extremis*, the "smoking gun" to justify impeachment). The general public seems to judge a president, at least in part, by what its members think is happening to them, in their own lives: congressmen, officials, interest groups appear to judge by what they guess, at given times, he can do for or to their causes. Members of the press interpret both of these and spread a simplified criterion affecting both, the legislative box score, a standard of the press's own devising. The White House denigrates them all except when it does well.

6. Implementation. The corporate chief, supposedly, does more than choose a strategy and set a course of policy; he also is supposed to oversee what happens after, how in fact intentions turn into results, or if they don't to take corrective action, monitoring through his information system, and acting, if need be, through his personnel system. A president, by contrast, while himself responsible for budgetary proposals, too, in many spheres of policy appears ill-placed and ill-equipped to monitor what agencies of states, of cities, corporations, unions, foreign governments are up to or to change personnel in charge. Yet these are very often the executants of "his" programs. Apart from defense and diplomacy the federal government does two things in the main: it issues and applies regulations and it awards grants in aid. Where these are discretionary, choice usually is vested by statute in a Senate-confirmed official well outside the White House. Monitoring is his function, not the president's except at second hand. And final action is the function of the subjects of the rules and funds; they mostly are not federal personnel at all. In defense, the arsenals and shipyards are gone; weaponry comes from the private sector. In foreign affairs it is the *other* governments whose actions we would influence. From implementors like these a president is far removed most of the time. He intervenes, if at all, on a crash basis, not through organization incentives.

Underlying these lists' sharpest distinctions between public and private management is a fundamental *constitutional difference*. In business, the functions of general management are centralized in a single individual: the chief executive officer. The goal is authority commensurate with responsibility. In contrast, in the U.S. government, the functions of general management are constitutionally spread among competing institutions: the executive, two houses of Congress, and the courts. The constitutional goal was "not to promote efficiency but to preclude the exercise of arbitrary power," as Justice Brandeis observed. Indeed, as *The Federalist Papers* makes starkly clear, the aim was to create incentives to compete: "the great security against a gradual concentration of the several powers in the same branch, consists in giving those who administer each branch the constitutional means and personal motives to resist encroachment of the others. Ambition must be made to counteract ambition."[12] Thus, the general management functions concentrated in the CEO of a private business are, by constitutional design, spread in the public sector among a number of competing institutions and thus shared by a number of individuals whose ambitions are set against one another. For most areas of public policy today, these individuals include at the federal level the chief elected official, the chief appointed executive, the chief career official, and several congressional chieftains. Since most public services are actually delivered by state and local governments, with independent sources of authority, this means a further array of individuals at these levels.

AN OPERATIONAL PERSPECTIVE: HOW ARE THE JOBS & RESPONSIBILITIES OF DOUG COSTLE, DIRECTOR of EPA, & ROY CHAPIN, CEO of AMERICAN MOTORS, SIMILAR AND DIFFERENT?

If organizations could be separated neatly into two homogeneous piles, one public and one private, the task of identifying similarities and differences between managers of these enterprises would be relatively easy. In fact, as Dunlop has pointed out, "the real world of management is composed of distributions, rather than single undifferentiated forms, and there is an increasing variety of hybrids." Thus for each major attribute of organizations, specific entities can be located on a spectrum. On most dimensions, organizations classified as "predominantly public" and those "predominantly private" overlap.[13] Private business organizations vary enormously among themselves in size, in management structure and philosophy, and in the constraints under which they operate. For example, forms of ownership and types of managerial control may be somewhat unrelated. Compare a family-held enterprise, for instance, with a public utility and a decentralized conglomerate, a Bechtel with ATT and Textron. Similarly, there are vast differences in management of governmental organizations. Compare the Government Printing Office or TVA or the police department of a small town with the Department of Energy or the Department of Health and Human Services. These distributions and varieties should encourage penetrating comparisons within both business and governmental organizations, as well as contrasts and comparisons across these broad categories, a point to which we shall return in considering directions for research.

Absent a major research effort, it may nonetheless be worthwhile to examine the jobs and responsibilities of two specific managers, neither polar extremes, but one clearly public, the other private. For this purpose, and primarily because of the availability of cases that describe the problems and opportunities each confronted, consider Doug Costle, Administrator of EPA, and Roy Chapin, CEO of American Motors.[14]

DOUG COSTLE, ADMINISTRATOR OF EPA, JANUARY 1977

The mission of EPA is prescribed by laws creating the agency and authorizing its major programs. That mission is "to control and abate pollution in the areas of air, water, solid wastes, noise, radiation, and toxic substances. EPA's mandate is to mount an integrated, coordinated attack on environmental pollution in cooperation with state and local governments."[15]

EPA's organizational structure follows from its legislative mandates to control particular pollutants in specific environments: air and water, solid wastes, noise, radiation, pesticides, and chemicals. As the new administrator, Costle inherited the Ford administration's proposed budget for EPA of $802 million for federal 1978 with a ceiling of 9,698 agency positions.

The setting into which Costle stepped is difficult to summarize briefly. As Costle characterized it:

> "Outside there is a confusion on the part of the public in terms of what this agency is all about; what it is doing, where it is going."
>
> "The most serious constraint on EPA is the inherent complexity in the state of our knowledge, which is constantly changing."
>
> "Too often, acting under extreme deadlines mandated by Congress, EPA has announced regulations, only to find out that they knew very little about the problem. The central problem is the inherent complexity of the job that the agency has been asked to do and the fact that what it is asked to do changes from day to day."
>
> "There are very difficult internal management issues not amenable to a quick solution: the skills mix problems within the agency; a research program with laboratory facilities scattered all over the country and cemented in place, largely by political alliances on the Hill that would frustrate efforts to pull together a coherent research program."
>
> "In terms of EPA's original mandate in the bulk pollutants we may be hitting the asymptotic part of the curve in terms of incremental clean-up costs. You have clearly conflicting national goals: energy and environment, for example."

Costle judged his six major tasks at the outset to be:

- Assembling a top management team (six assistant administrators and some 25 office heads).
- Addressing EPA's legislative agenda (EPA's basic leg-

islative charter – the Clean Air Act and the Clean Water Act – was being rewritten as he took office; the pesticides program was up for reauthorization also in 1977).

- Establishing EPA's role in the Carter Administration (aware that the Administration would face hard tradeoffs between the environment and energy, energy regulations and the economy, EPA regulations of toxic substances and the regulations of FDA, CSPS, and OSHA. Costle identified the need to build relations with the other key players and to enhance EPA's standing).

- Building ties to constituent groups (both because of their role in legislating the agency's mandate and in successful implementation of EPA's programs).

- Making specific policy decisions (for example, whether to grant or deny a permit for the Seabrook Nuclear Generating Plant cooling system. Or how the Toxic Substance Control Act, enacted in October 1976, would be implemented; this act gave EPA new responsibilities for regulating the manufacture, distribution, and use of chemical substances so as to prevent unreasonable risks to health and the environment. Whether EPA would require chemical manufacturers to provide some minimum information on various substances, or require much stricter reporting requirements for the 1,000 chemical substances already known to be hazardous, or require companies to report all chemicals, and on what timetable, had to be decided and the regulations issued).

- Rationalizing the internal organization of the agency (EPA's extreme decentralization to the regions and its limited technical expertise).

No easy job.

ROY CHAPIN AND AMERICAN MOTORS, JANUARY 1967

In January 1967, in an atmosphere of crisis, Roy Chapin was appointed Chairman and Chief Executive Officer of American Motors (and William Luneburg, President and Chief Operating Officer). In the four previous years, AMC unit sales had fallen 37 percent and market share from over 6 percent to under 3 percent. Dollar volume in 1967 was off 42 percent from the all-time high of 1963 and earnings showed a net loss of $76 million on sales of $656 million. Columnists began writing obituaries for AMC. *Newsweek* characterized AMC as "a flabby dispirited company, a product solid enough but styled with about as much flair as corrective shoes, and a public image that melted down to one unshakable label: loser." Said Chapin, "We were driving with one foot on the accelerator and one foot on the brake. We didn't know where...we were."

Chapin announced to his stockholders at the outset that "we plan to direct ourselves most specifically to those areas of the market where we can be fully effective. We are not going to attempt to be all things to all people, but to concentrate on those areas of consumer needs we can meet better than anyone else." As he recalled, "There were problems early in 1967 which demanded immediate attention, and which accounted for much of our time for several months. Nevertheless, we began planning beyond them, establishing objectives, programs and timetables through 1972. Whatever happened in the short run, we had to prove ourselves in the marketplace in the long run."

Chapin's immediate problems were five:

- The company was virtually out of cash and an immediate supplemental bank loan of $20 million was essential.

- Car inventories – company owned and dealer owned – had reached unprecedented levels. The solution to this glut took five months and could be accomplished only by a series of plant shutdowns in January 1967.

- Sales of the Rambler American series had stagnated and inventories were accumulating: a dramatic merchandising move was concocted and implemented in February, dropping the price tag on the American to a position midway between the VW and competitive smaller U.S. compacts, by both cutting the price to dealers and trimming dealer discounts from 21 percent to 17 percent.

- Administrative and commercial expenses were judged too high and thus a vigorous cost reduction program was initiated that trimmed $15 million during the first year. Manufacturing and purchasing costs were also trimmed significantly to approach the most effective levels in the industry.

- The company's public image had deteriorated: the press was pessimistic and much of the financial community had written it off. To counteract this, numerous formal and informal meetings were held with bankers, investment firms, government officials, and the press.

As Chapin recalls, "With the immediate fires put out, we could put in place the pieces of a corporate growth plan – a definition of a way of life in the auto industry for American Motors. We felt that our reason for being, which would enable us not just to survive but to grow, lay in bringing a different approach to the auto market – in picking our spots and then being innovative and aggressive." The new corporate growth plan included a dramatic change in the approach to the market to establish a "youthful image" for the company (by bringing out new sporty models like the Javelin and by entering the racing field), "changing the product line from one end to the other" by 1972, acquiring Kaiser Jeep (selling the company's non-transportation assets and concentrating on specialized transportation, including Jeep, a company that had lost money in each of the preceding five years but that Chapin believed could be turned around by substantial cost reductions and economies of scale in manufacturing, purchasing, and administration).

Chapin succeeded for the year ending September 30, 1971. AMC earned $10.2 million on sales of $1.2 billion.

Recalling the list of general management functions in Table 2, which similarities and differences appear salient and important?

Strategy

Both Chapin and Costle had to establish objectives and priorities and to devise operations plans. In business, "corporate strategy is the pattern of major objectives, purposes, or goals and essential policies and plans for achieving those goals, stated in such a way as to define what business the company is in or is to be in and the kind of company it is or is to be."[16] In reshaping the strategy of AMC and concentrating on particular segments of the transportation market, Chapin had to consult his board and had to arrange financing. But the control was substantially his.

How much choice did Costle have at EPA as to the "business it is or is to be in" or the kind of agency "it is or is to be"? These major strategic choices emerged from the legislative process which mandated whether he should be in the business of controlling pesticides or toxic substances and if so on what timetable, and occasionally, even what level of particulate per million units he was required to control. The relative role of the president, other members of the administration (including White House staff, congressional relations, and other agency heads), the EPA Administrator, congressional committee chairmen, and external groups in establishing the broad strategy of the agency constitutes an interesting question.

Managing Internal Components

For both Costle and Chapin, staffing was key. As Donald Rumsfeld has observed, "the single most important task of the chief executive is to select the right people. I've seen terrible organization charts in both government and business that were made to work well by good people. I've seen beautifully charted organizations that didn't work very well because they had the wrong people."[17]

The leeway of the two executives in organizing and staffing were considerably different, however. Chapin closed down plants, moved key managers, hired and fired, virtually at will. As Michael Blumenthal has written about Treasury, "If you wish to make substantive changes, policy changes, and the Department's employees don't like what you're doing, they have ways of frustrating you or stopping you that do not exist in private industry. The main method they have is Congress. If I say I want to shut down a particular unit or transfer the function of one area to another, there are ways of going to Congress and in fact using friends in the Congress to block the move. They can also use the press to try to stop you. If I at Bendix wished to transfer a division from Ann Arbor to Detroit because I figured out that we could save money that way, as long as I could do it decently and carefully, it's of no lasting interest to the press. The press can't stop me. They may write about it in the local paper, but that's about it."[18]

For Costle, the basic structure of the agency was set by law. The labs, their location, and most of their personnel were fixed. Though he could recruit his key subordinates, again restrictions like the conflict of interest laws and the prospect of a Senate confirmation fight led him to drop his first choice for the Assistant Administrator for Research and Development, since he had worked for a major chemical company. While Costle could resort to changes in the process for developing policy or regulations in order to circumvent key office directors whose views he did not share, for example, Eric Stork, the Deputy Assistant Administrator in charge of Mobile Source Air Program, such maneuvers took considerable time, provoked extensive infighting, and delayed significantly the development of Costle's program.

In the direction of personnel and management of the personnel system, Chapin exercised considerable authority. While the United Auto Workers limited his authority over workers, at the management level he assigned people and reassigned responsibility consistent with his general plan. While others may have felt that his decisions to close down particular plants or to drop a particular product were mistaken, they complied. As George Shultz has observed: "One of the first lessons I learned in moving from government to business is that in business you must be very careful when you tell someone who is working for you to do something because the probability is high that he or she will do it."[19]

Costle faced a civil service system designed to prevent spoils as much as to promote productivity. The Civil Service Commission exercised much of the responsibility for the personnel function in his agency. Civil service rules severely restricted his discretion, took long periods to exhaust, and often required complex maneuvering in a specific case to achieve any results. Equal opportunity rules and their administration provided yet another network of procedural and substantive inhibitions. In retrospect, Costle found the civil service system a much larger constraint on his actions and demand on his time than he had anticipated.

In controlling performance, Chapin was able to use measures like profit and market share, to decompose those objectives to subobjectives for lower levels of the organization and to measure the performance of managers of particular models, areas, divisions. Cost accounting rules permitted him to compare plants within AMC and to compare AMC's purchases, production, and even administration with the best practice in the industry.

Managing External Constituencies

As chief executive officer, Chapin had to deal only with the Board. For Costle, within the executive branch but beyond his agency lay many actors critical to the achievement of his agency objectives: the president and the White House, Energy, Interior, the Council on Environmental Quality, OMB. Actions each could take, either independently or after a process of consultation in which they disagreed with him, could frustrate his agency's achievement of its assigned mission. Consequently, he spent considerable time building his agency's reputation and capital for interagency disputes.

Dealing with independent external organizations was a necessary and even larger part of Costle's job. Since his agency, mission, strategy, authorizations, and appropriations emerged from the process of legislation, attention to congressional committees, congressmen, congressmen's staff, and people who affect congressmen and congressional staffers rose to the top of Costle's agenda. In the first year, top-level EPA officials appeared over 140 times before some 60 different committees and subcommittees.

Chapin's ability to achieve AMC's objectives could also be affected by independent external organization: competitors, government (the Clean Air Act that was passed in 1970), consumer groups (recall Ralph Nader), and even suppliers of oil. More than most private managers, Chapin had to deal with the press in attempting to change the image of AMC. Such occasions were primarily at Chapin's initiative and around events that Chapin's public affairs office orchestrated, for example, the announcement of a new racing car. Chapin also managed a marketing effort to persuade consumers that their tastes could best be satisfied by AMC products.

Costle's work was suffused by the press: in the daily working of the organization, in the perception by key publics of the agency and thus the agency's influence with relevant parties, and even in the setting of the agenda of issues to which the agency had to respond.

For Chapin, the bottom line was profit, market share, and the long-term competitive position of AMC. For Costle, what are the equivalent performance measures? Blumenthal answers by exaggerating the difference between appearance and reality: "At Bendix, it was the reality of the situation that in the end determined whether we succeeded or not. In the crudest sense, this meant the bottom line. You can dress up profits only for so long – if you're not successful, it's going to be clear. In government there is no bottom line, and that is why you can be successful if you appear to be successful – though, or course, appearance is not the only ingredient of success."[20] Rumsfeld says, "In business, you're pretty much judged by results. I don't think the American people judge government officials this way...In government, too often you're measured by how much you seem to care, how hard you seem to try – things that do not necessarily improve the human condition...It's a lot easier for a President to get into something and end up with a few days of good public reaction than it is to follow through, to pursue policies to a point where they have a beneficial effect on human lives."[21] As George Shultz says, "In government and politics, recognition and therefore incentives go to those who formulate policy and maneuver legislative compromise. By sharp contrast, the kudos and incentives in business go to the persons who can get something done. It is execution that counts. Who can get the plant built, who can bring home the sales contract, who can carry out the financing, and so on."[22]

This casual comparison of one public and one private manager suggests what could be done if the issue of comparisons were pursued systematically, horizontally across organizations and at various levels within organizations. While much can be learned by examining the chief executive officers of organizations, still more promising should be comparisons among the much larger numbers of middle managers. If one compared, for example, a regional administrator of EPA and an AMC division chief, or two comptrollers, or equivalent plant managers, some functions would appear more similar, and other differences would stand out. The major barrier to such comparisons is the lack of cases describing problems and practices of middle-level managers.[23] This should be a high priority in further research.

The differences noted in this comparison, for example, in the personnel area, have already changed with the Civil Service Reform Act of 1978 and the creation of the Senior Executive Service. Significant changes have also occurred in the automobile industry: under current circumstances, the CEO of Chrysler may seem much more like the administrator of EPA. More precise comparison of different levels of management in both organizations, for example, accounting procedures used by Chapin to cut costs significantly as compared to equivalent procedures for judging the costs of EPA mandated pollution control devices, would be instructive...

NOTES

1. Selma J. Mushkin, Frank H. Sandifer, and Sally Familton, Current Status of Public Management Research Conducted by or Supported by Federal Agencies (Washington, D.C.: Public Services Laboratory, Georgetown University, 1978), p. 10.

2. Ibid., p. 11.

3. Though frequently identified as the author who established the complete separation between "policy" and "administration," Woodrow Wilson has in fact been unjustly accused. "It is the object of administrative study to discover, first, what government can properly and successfully do, and, secondly, how it can do these proper things with the utmost possible efficiency..." (Wilson, "The Study of Public Administration," published as an essay in 1888 and reprinted in Political Science Quarterly, December 1941, p. 481). For another statement of the same point, see Brooks Adams, The Theory of Social Revolutions (Macmillan, 1913), pp. 207-208.

4. See Dwight Waldo, "Organization Theory: Revisiting the Elephant," PAR (November-December 1978). Reviewing the growing volume of books and articles on organization theory, Waldo notes that "growth in the volume of the literature is not to be equated with growth in knowledge."

5. See Cases in Public Policy and Management, Spring 1979, of the Intercollegiate Case Clearing House for a bibliography containing descriptions of 577 cases by 366 individuals from 79 institutions. Current casework builds on and expands earlier efforts on the Inter-University Case Program. See, for example, Harold Stein, ed., Public administration and Policy Development: A Case Book (Orland, Fla.: Harcourt Brace Jovanovich, 1952), and Edwin A. Bock and Alan K. Campbell, eds., Case Studies in American Government (Englewood Cliffs, N.J.: Prentice-Hall, 1962).

6. Luther Gulick and Al Urwick, eds., Papers in the Science of Public Administration (Washington, D.C.: Institute of Public Administration, 1937).

7. See, for example, Chester I. Barnard, The Functions of the Executive (Cambridge, Mass.: Harvard University Press, 1938), and Peter F. Drucker, Management Tasks, Responsibilities, Practices (New York: Harper & Row, 1974). Barnard's recognition of human relations added an important dimension neglected in earlier lists.

8. See, for example, "A Businessman in a Political Jungle," Fortune (April 1964); "Candid Reflections of a Businessman in Washington," Fortune (January 29, 1979); "A Politician Turned Executive," Fortune (September 10, 1979); and "The Abrasive Interface," Harvard Business Review (November-December 1979) for the views of Romney, Blumenthal, Rumsfeld, and Shultz, respectively.

9. John T. Dunlop, "Public Management," draft of an unpublished paper and proposal, Summer 1979.

10. Hal G. Rainey, Robert W. Backoff, and Charles N. Levine, "Comparing Public and Private Organizations," Public Administration Overview (March-April 1976).

11. Richard E. Neustadt, "American Presidents and Corporate Executives," paper prepared for a meeting of the National Academy of Public Administration's Panel on Presidential Management, October 7-8, 1979.

12. Clinton Rossiter, ed., The Federalist Papers (New York: New American Library, 1961), No. 51. The word department has been replaced by branch, which was its meaning in the original papers.

13. Failure to recognize the fact of distributions has led some observers to leap from one instance of similarity between public and private to general propositions about similarities between public and private institutions or management. See, for example, Michael Murray, "Comparing Public and Private Management: An Exploratory Essay," Public Administration Review (July-August, 1975).

14. These examples are taken from Bruce Scott, "American Motors Corporation" (Intercollegiate Case Clearing House #9-364-001); Charles B. Weigle with the collaboration of C. Roland Christensen, "American Motors Corporation II" (intercollegiate Case Clearing House #6-372-359); Thomas R. Hitchner and Jacob Lew under the supervision of Philip B. Heymann and Stephen B. Hitchener, "Douglas Costle and the EPA (A)" (Kennedy School of Government Case #C94-78-216), and Jacob Lew and Stephen B. Hitchner, "Douglas Costle and the EPA (B)" (Kenney School of Government Case #C96-78-217). For an earlier exploration of a similar comparison, see Joseph Bower, "Effective Public Management," Harvard Business Review (March-April, 1977).

15. U.S. Government Manual, 1978/1979, p. 507.

16. Kenneth R. Andrews, The Concepts of Corporate Strategy (New York: Dow-Jones-Irwin, 1971), p. 28.

17. "A Politician-Turned-Executive," Fortune (September 10, 1979), p. 92.

18. "Candid Reflections of a Businessman in Washington," Fortune (January 29, 1979), p. 39.

19. "The Abrasive Interface," Harvard Business Review (November-December 1979), p. 95.

20. Fortune (January 29, 1979), p. 36.

21. Fortune (September 10, 1979), p. 90.

22. Harvard Business Review (November-December 1979), p. 95.

23. The cases developed by Boston University's Public Management Program offer a promising start in this direction.

ABOUT THE AUTHOR

Graham Allison is the "founding dean" of Harvard University's Kennedy School of Government. He is Director of the Belfer Center for Science and International Affairs and the Douglas Dillon Professor of Government, at Harvard. A recipient of the Department of Defense's highest civilian award, Dr. Allison served at the Pentagon in the Reagan and Clinton administrations.

From: Graham T. Allison, Jr., "Public and Private Management: Are They Fundamentally Alike in All Unimportant Respects?," *Setting Public Management Research Agendas: Integrating the Sponsor, Producer, and User*, OPM Document 127-53-1 (February 1980): 27-38. Used with permission.

14

CHAPTER 14

AIRPOWER AS STRATEGIC LABORATORY

This chapter highlights the unique strategic nature of American airpower, tracing its evolution over the last century and examining the changing role it plays in national security. Airpower is a broad concept, as described by Billy Mitchell when he wrote "Air power is the ability to do something in or through the air," and encompasses not only military power but civil and industrial might.

INTRODUCTION

The readings presented here concentrate on the Air Force definition of airpower as "the ability to project military power or influence through the control and exploitation of air, space, and cyberspace." The historical development of airpower provides a useful case study in strategic leadership by analyzing how the US Air Force successfully evolved over time as a result of visionary strategic leadership.

General Carl Spaatz was a brilliant combat leader who played a central role in the establishment of the US Air Force as an independent Service, separate and equal to the Army and Navy. His treatise "Strategic Air Power: Fulfillment of a Concept" sets the stage by validating the importance of a well-thought-out strategy in achieving one's objectives. In presenting his argument, General Spaatz asserts that Germany lost World War II due in part to its misuse of airpower and he warns of the importance of maintaining a strong and prepared Air Force in peacetime.

The second article, "Warden and the Air Corps Tactical School: What Goes Around Comes Around," presents two striking examples of airpower theorists relying upon their unique strategic perspectives to conceptualize the battlespace and apply a systems approach to strategy. This article builds on the systems thinking article you read in chapter 12. Through contrast and comparison the author identifies similarities, strengths, and shortcomings of the Air Corps Tactical School's ideas promoted throughout the 1930s and those of Colonel John Warden's Five-Ring Theory published in the late 1980s.

Having surveyed the history of the US Air Force as an independent service in the first two articles, we will next turn our attention to how airpower is being redefined today. In "Cyberspace: The New Air and Space?" the author explores the cyber domain and the important role information technology plays in national security. Today's airpower advocates view cyberspace as a natural complement to the traditional airpower mediums of air and space used to project military power.

Our fourth reading comes directly from Air Force doctrine, and in it one can trace the influences of the transformational airpower leaders of the past, great captains like Hap Arnold, Jimmy Doolittle, and Tooey Spaatz, as well as less familiar visionaries such as George Kenney, William Tunner, and Alexander de Seversky. The selected excerpts from AFDD 1 summarize how the US Air Force role has expanded over time to incorporate twelve core functions that embody what the world's most powerful military force manifests across the range of military operations.

The final reading raises a challenging issue for current and future air-minded leaders. For decades the US has relied upon a nuclear triad of land-based intercontinental ballistic missiles (ICBMs), submarine-launched ballistic missiles (SLBMs), and bomber aircraft to provide deterrence for our nation and its allies. Some have called upon the US to divest itself of these expensive and complex weapon systems on a path toward nuclear disarmament. In his article "Should the United States Maintain the Nuclear Triad?" Dr. Adam Lowther examines this debate and its strategic implications.

CHAPTER OUTLINE

This chapter's readings are:

Strategic Air Power: Fulfillment of a Concept
Gen Carl Spaatz, "Strategic Air Power: Fulfillment of a Concept," Foreign Affairs 24 (1945/1946): 385-396.

Warden & The Air Corps Tactical School: What Goes Around Comes Around
Maj Howard D. Belote, "Warden and the Air Corps Tactical School: What Goes Around Comes Around," Airpower Journal (Fall 1999): 39-47.

Cyberspace: The New Air & Space?
Lt Col David A. Umphress, "Cyberspace: The New Air and Space?," Air & Space Power Journal (Spring 2007): 50-55.

Air Force Basic Doctrine
US Air Force, AFDD-1, *Air Force Basic Doctrine,* (Maxwell AFB, AL: Air University, October 2011).

Should the US Maintain the Nuclear Triad?
Dr. Adam B. Lowther, "Should the United States Maintain the Nuclear Triad?," Air & Space Power Journal (Summer 2010): 23-29.

CHAPTER GOALS

1. Summarize the evolution of American airpower over the last century.
2. Appreciate the role of information, space, and cyberspace in modern airpower theory.
3. Develop an understanding of the Air Force's core functions.

14.1 Strategic Air Power: Fulfillment of a Concept

By General Carl Spaatz

OBJECTIVES:

1. List Germany's three critical mistakes in its use of airpower, as determined by Gen Spaatz.
2. Define the term "strategic bombing."
3. Name the three principles of combat that strategic bombing takes advantage of.
4. Describe, in your own terms, five lessons that the United States learned from the use of strategic air power in World War II.

World War II might have ended differently had our enemies understood and made correct use of Strategic Air Power.

In the elation of victory it is well for us to remember the year 1942 when the conquests of the Axis Powers reached their apogee. Europe was a Nazi fortress, mined and ribbed with the latest improvements in surface defense, over which the Luftwaffe reigned supreme. In the German view, science had made that fortress impregnable. Astonishing feats of logistics had enabled the Wehrmacht to stretch from the Pyrenees to the Volga and the Caucasus; and Italian contingent armies in North Africa approached the Nile. Japan also was a fortress; and outside it, the Japanese reach extended from Burma in a vast arc to the Aleutians.

The outlook for the Allies was grim. By all time-tested and "proven" methods of warfare the combined might of the Axis Powers seemed unconquerable. Their resources in manpower and materiel were such that they could ward off exhaustion for an indefinite period of time. Sea blockade, therefore, could not be counted on to have the strangling effect it produced in World War I. Our land and sea forces, supported by air, could be expected to contain the most advanced echelons of our enemies, and gradually to drive back their main armies into their heavily fortified citadels. But the essential question remained. How was their military power to be crushed behind their ramparts without undertaking an attritional war which might last years, which would cost wealth that centuries alone could repay and which would take untold millions of lives? The man in the street asked, with reason: "How can we ever beat them? With what?"

The development of a new technique was necessary. Some new instrument had to be found, something untried and therefore "unproven," something to "spark the way" to early and complete victory. The outcome of the total war hung in the balance until that new technique had been found and proved decisive in all-out assault. The new instrument was Strategic Air Power. In 1942 it was already in the process of development.

II. THE GERMAN STRATEGIC FAILURE, 1940

The effectiveness of the new technique had been given negative demonstration by Germany's history-making mistake in 1940. After Dunkirk, Hitler stood on the threshold of his goal, the domination of all Europe. Which way would he strike next? France was prostrate; Spain was not unfriendly. Two trained German parachute divisions were on the alert to drop on Gibraltar, the capture of which would have corked up the western exit of the British Mediterranean fleet. The war on Britain's life stream of shipping could then have been increased to unbearable intensity. On the other hand, there, just across the Channel, lay Britain, without the thousand new field guns which the B.E.F. had left behind in Belgium. Guarding the narrow strip of water were powerful elements of the British Navy, and an unknown number of British fighter airplanes. Hitler made his choice: it was to let Gibraltar wait, and to try for a "knock-out" blow against Britain from the air as a preliminary to turning on Russia. It was his historic opportunity, which was never to return.

Fortunately for us, neither Hitler nor the German High Command understood the strategic concept of air power or the primary objective of a strategic air offensive. The Germans had air supremacy on the Continent. They also had air superiority in numbers over Britain; but they were unable to establish control of the air, and this was essential to carry out sustained operations. The German bombers were lightly armed. The German fighters were used in close support of the bombers. The British had the surprise of radar and eight-gunned fighters. Technically and tactically the R.A.F. was superior. Air control can be established by superiority in numbers, by better employment, by better equipment, or by a combination of these factors. The Germans might have gained control of the air if their fighters had been used in general support instead of close support of the bombers, or if their bombers had done more accurate and effective bombing (*e.g.* on the British airfields), or if all the German air force had been directed against Britain.

It was apparent to observers in 1940[1] that the German leadership was wedded to the old concept that air power was restricted to support of fast-moving ground troops and that it did not have an independent mission of its own. This tactical concept had been successfully implemented against Poland and France by the Stuka-Panzer combination, under conditions of German air supremacy. The bombing of Britain, on the other hand, was a strategic task, for the successful accomplishment of which German control of the air first had to be established. The Germans disregarded this absolute necessity. First, they had not built heavy bombers which could carry enough armament to be relatively secure. The lightly-armed Ju 88's, He III ' s and Do 17's which carried the bombs were no match for the British eight-gunned fighters, aided by the warnings of secret radar. They were shot down in swarms. Second, the German fighters outnumbered the R.A.F. Hurricanes and Spitfires. Their proper function was to destroy R.A.F. fighters. Instead, they kept close formation to cover the inadequately armed bombers — a defensive role which could never win control of the air.

Viewed historically, the German failure in the Blitz demonstrated the wrong technique for strategic bombing. The German mistakes were: 1, inadequate armament on the bombers; 2, no capability for precision bombing; 3, use of the fighters in close support of the bombers instead of in general support.

Germany had the industrial capacity and skill to build properly armed heavy bombers before and during the early years of the war. The four-engined Focke-Wulfe was in operation, but was used against shipping from Norway and France. The He 177, with two propellers on four motors, was a failure, and wasted two years of effort. Consequently, the Luftwaffe attempted the strategic reduction of Britain from the air with means which could have been successful only through the proper use of German fighter superiority. But the Nazi war leaders (to whom the Luftwaffe was completely subservient, which meant that independent air thinking was in abeyance) did not grasp the strategic concept. If they had understood it, and had built heavy well-armed bombers, and had used their fighters to gain control of the air, they could actually have reduced Britain to a shambles in 1940. Later, by applying the strategic lessons, they probably would have been able to hold the line of the Volga by bombing Russian war plants in the Urals and beyond. Once the success of strategic air warfare had been demonstrated, it is conceivable that Hitler would not have declared war on America when he did. In any case, we would have been too late for this particular war, and we would have been deprived of the use of the United Kingdom as a base when the time came for us to fight.

The historic penalty paid by the Nazis for their mistake was that they have passed into oblivion and Germany lies in ruins.

III. THE STRATEGIC CONCEPT: THE IDEA AND THE WEAPON

Strategic bombing, the new technique of warfare which Germany neglected in her years of triumph, and which Britain and America took care to develop, may be defined as being an independent air campaign, intended to be decisive, and directed against the essential war-making capacity of the enemy. Its immeasurable advantage over two-dimensional techniques is that its units (heavy bombers and fighter escorts) are not committed to position in battle; on the contrary, they carry out their assigned missions, and then return to base to prepare for fresh assault.

What makes strategic bombing the most powerful instrument of war thus far known is its effective application of:

1. The principle of mass, by its capacity to bring all its forces from widely distributed bases simultaneously to focus on single targets. Such concentration of combat power has never been possible before.

2. The principle of objective, by its capacity to select for destruction those elements which are most vital to the enemy's war potential, and to penetrate deep into the heart of the enemy country to destroy those vital elements wherever they are to be found. These main objectives, reached during hostilities by strategic bombing following the establishment of control of the air, have not been attained historically by surface forces until toward the end of field campaigns.

3. The principle of economy of force, by its capacity to concentrate on a limited number of vital target systems instead of being compelled to disperse its force on numerous objectives of secondary importance, and by its capacity to select for destruction that portion of a target system which will yield the desired effect with the least expenditure of force.

Strategic bombing is thus the first war instrument of history capable of stopping the heart mechanism of a great industrialized enemy. It paralyzes his military power at the core. It has a strategy and tactic of mobility and flexibility which are peculiar to its own medium, the third dimension. And it has a capacity, likewise peculiar, to carry a tremendous striking force, with unprecedented swiftness, over the traditional line of war (along which the surface forces are locked in battle on land and sea) in order to destroy war industries and arsenals and cities, fuel plants and supplies, transport and communications — in fact, the heart and the arteries of war economy — so that the enemy's will to resist is broken through nullification of his means.

British air leaders had this strategic concept in mind at the beginning of the war. But they lacked the means to carry it out. Their daylight raids on German industrial targets in 1940 resulted in prohibitive losses. Accordingly, the R.A.F. turned to night bombing, which was feasible despite the Luftwaffe's air supremacy over Germany because effective night fighters had not yet appeared. The British developed the most effective heavy night bomber, the Lancaster, which went into action in 1943 and remained the greatest load-carrier of the air war in Europe.

The strategic concept had also been the focus of studies and planning in the United States Army Air Forces in the 1930's. The American version was built around the B-17 for precision bombing by daylight. Daylight bombing was still regarded with skepticism in some quarters because of the German experience in the 1940 Blitz and the British experience over German targets. Both our weapon and our organization remained untried. It was feared that the losses in daylight bombing would be prohibitive. Accordingly, there was an inclination on the part of experienced war leaders to put all Allied strategic bombers on the night run.

The critical moment in the decision whether or not this should be done came on January 21, 1943. On that date the Combined Chiefs of Staff finally sanctioned continuance of bombing by day and issued the Casablanca directive which called for the "destruction and dislocation of the German military industrial and economic system and the undermining of the morale of the German people to the point where their capacity for armed resistance is fatally weakened." To implement this directive there was drawn up a detailed plan, "The Combined Bomber Offensive Plan," which was approved by the Combined Chiefs of Staff, June 10, 1943, and issued to British and American air commanders. Strategic bombing at last had the green light; and it possessed a plan of operations of its own, with an approved order of priorities in targets, to achieve the objectives of the Casablanca directive. That plan called for bombing by night and by day, round the clock.

IV. FULFILLMENT OF THE CONCEPT

As far back as the time of Pearl Harbor the Army Air Forces had the Idea; but the Idea still remained to be worked out by experiment in the grim practice of war. In order to do this we first had to "forge" the weapon, develop the proper technique to make it decisive in battle, prepare the necessary bases within operational range of the proposed targets, and then establish control of the air before proceeding to the all-out assault. All these things took time. The building of the Air Forces with sufficient striking power to carry out the strategic tasks, as ultimately outlined in the Combined Bomber Offensive Plan, required a national effort of unprecedented magnitude, and two and a half years of time. Those years were provided by the unwavering resistance of our Allies to our common enemies.

It took time to "forge" the weapon. The portion of America's industrial power devoted to the manufacture of airplanes and their equipment had already been stepped up by British and French war orders. This capacity was shifted to fulfillment of our own needs. Constant technical research made for improved designs and for modifications, based on experience in battle, to arrive at an all-weather weapon capable of self-defense. At the peak of our strength, in 1944, there were nearly 80,000 airplanes of all types under the control of the A.A.F., of which more than half were in combat. The heavy bombers, the B-17's and the B-24's, along with the fighters (P-51, P-47, and P-38) which provided the long-range escort beginning in the autumn of 1943, accomplished the decisive strategic task in Europe. The B-29, the most powerful airplane ever built, accompanied by the P-51, was equally decisive in destroying Japan's capacities to wage war. The quantity production of the heavy bomber in three types and of the necessary long-range fighter escorts was an achievement which will stand to the historic credit of America's industrial genius in support of air power.

It took time to acquire a new technique for the effective employment of the chosen weapon. There never had been a strategic air war on the scale projected. The proper methods had to be learned by experiment. The Army Air Force, which had 1,300 flying officers of the Regular Army on active duty in 1940, expanded to reach a total of 2,300,000 personnel in 1944. Technical training was necessary in the organization of air and ground crews (the backbone of an air force) to man the 220 groups projected, as well as in intelligence and target selection, in communications, weather, radio, radar, tactical air doctrine, etc. Gradual mastery of the new technique kept pace with production of the weapons.

It took time to prepare bases within operational range of the enemy's vital war potentials, and to build up the supply system and the supplies necessary to sustain operations. In its global war the A.A.F. needed bases in such widely distributed theaters that the allocation of materiel was a constant problem. The European theater was given top priority in airplanes, but circumstances at times dictated diversions to the Pacific. The base in the United Kingdom had to be established in spite of the enemy submarine menace in 1942. The "Torch" operation in North Africa in November 1942 depleted the Eighth Air Force, both as to airplanes and personnel, but led one year later to the creation of a second strategic base in Italy. The activation of the Fifteenth Air Force in Italy in November 1943 made possible the coordination of bombing attacks from two theaters on the same German targets, thus implementing the principle of mass. In the Pacific, bases for the B-29's were first in China, and later were moved to the Marianas and Okinawa as the surface attack on Japanese forces closed in on Japan proper. The A.A.F. operational air bases around the world represented a triumph of American engineering ingenuity, whether by the laying of huge runways for the super-bombers, or by the conversion of swamps and deserts into air strips by means of steel mats.

Finally, it took time to gain control of the air, the absolutely necessary prerequisite for sustained strategic bombing. The German Air Force, although designed primarily to support ground troops, was a formidable defense — a fighting wall in the air. The task was to smash the wall, not only in order to clear the way for our heavy bombers over Germany, but also so as to remove the threat of air attack on our surface forces during and after the planned invasion. The duel with the German Air Force ensued.

In July 1943 an effort was made to get on with the first big task — the destruction of the German fighter system. These battles were a slugging match. A decision might have been forced if the Allies had had enough strength to continue beyond the one week of concentrated attack. During this period the line of battle was pushed back by whittling tactics of attrition from mid-Channel to the interior of Germany. Toward the end of 1943 there was at last sufficient force in hand. The long-range fighters needed to combat the enemy fighter defenses had been perfected, equipped with additional fuel tanks. Other equipment had likewise been modified under battle conditions. The Strategic Air Forces were ready to smash the German air wall, and then to proceed with the Combined Bomber Offensive.

On February 20, 1944, there began six days of perfect weather which were utilized for a continuous assault on the widely-dispersed German aircraft-frame factories and assembly plants. This sustained attack, called "The Big Week," fatally reduced the capabilities of the Luftwaffe. German aircraft production recovered; but the Allies retained control of the air throughout the remaining 14 months of hostilities.

In the minds of our air leaders the Big Week was the turning point in the war. That is, the success of the Big Week confirmed belief in the strategic concept. What had been in doubt was now a certainty. We knew now that we could destroy the German capacity to make war.

Having achieved control of the air, the Strategic Air Forces were employed on a twofold mission: 1, preparation for D-Day by the systematic destruction of the enemy's transport and communications; and 2, progressive destruction of his synthetic oil plants and other elements immediately vital to his continued resistance.

On April 16, 1945, the Headquarters of the U. S. Strategic Air Forces issued an order ending strategic bombing. The strategic air war in Europe was over; the concept had been fulfilled.

The lessons learned in the air war over Germany were applied with increasing vigor over Japan. The B-29 assault on the war industries in Japan proper began in the summer of 1944 with small attacks from China; these were augmented by attacks of similar weight from the Marianas beginning in November. The all-out mass offensive by the Twentieth Air Force began with the first low-level incendiary attack of March 9, 1945, and continued at accelerated frequency and intensity until Japan's capitulation on August 14, 1945. An invasion by the surface forces was not necessary. This air campaign will remain the classic prototype of the strategic concept as fulfilled in World War II.

V. APPRAISALS

The United States Strategic Bombing Survey, after nearly a year of study and six months of investigations in Germany, issued the following over-all judgment: "Allied air power was decisive in the war in Western Europe." Certain authoritative enemy judgments may be cited in support of this view.

The German reaction was well summed up by Lieutenant General Linnarz, Commander of the crack 26th Panzer Division, when he was interrogated on June 26, 1945, as follows:

> The basic conception of winning a war through strategic air power is sound. Historically, the strategic objective of any war has been to destroy the enemy's armies in the field. With increasing technological development, however, and the military fact that wars are no longer exclusively decided by generalship and battles, but by a nation's material might and war potential, it is obvious that in the future the first strategic objective in war cannot be the destruction of the armies in the field, but the destruction of the enemy's resources and war arsenals. Without these, the armies in the field are doomed to eventual defeat. A war might conceivably start with the attempt to destroy a nation's material power through employing a powerful weapon of long-range striking power. In this war, such a weapon was the long-range heavy bomber. In the future war it could conceivably be a type of perfected V-bomb.

In my opinion, you might have won the war through strategic bombing alone — granted adequate bases, tactically secured. Since you wanted to end the war quickly, you did not rely on strategic bombing alone; you fought the war in combined operations on land, sea and air. At the beginning of the war we failed to see that the material power of the coalition against us was strong enough to destroy our war industries by strategic air attacks, even if we took the whole Continent. As our leaders couldn't see this, and as you were unwilling to rely entirely on strategic bombing, you brought the war to an early and successful close by both strategic and tactical use of air power.

Professor Willi Messerschmitt, designer of the famous Me 109, 110, etc., stated when interrogated:

> One of the strategic mistakes was the failure to construct a fleet of long-range bombers to supplement submarine warfare in the Atlantic and thereby to deny the United States the ability to set up an operating air force within range of German industrial centers.

Albert Speer, Reich Minister for Armaments and War Production, said:

> The planned assaults on the chemical industry (synthetic oil) which began on May 13, 1944, caused the first serious shortages of indispensable basic products and therefore the greatest anxiety for the future conduct of the war. Actually, this type of attack was the most decisive factor in hastening the end of the war. . . . The attacks on the synthetic oil industry would have sufficed, without the impact of purely military events, to render Germany defenseless. Further targets of the same kind were to be found in the ball-bearing industry and in power stations. . . . The dispersal of important industries from west and northwest Germany to central and eastern Germany was carried out in 1942 and 1943. From 1944 onward, vital key industries were transferred to caves and other underground installations. Production was hindered not so much by these dispersals as by the shattering of transport and communication facilities. Consequently it can be said in conclusion that a bomb load is more effective if it is dropped upon economic targets than if it is expended upon towns and cities.

VI. LESSONS OF STRATEGIC AIR POWER

What are the chief lessons of our experience with the strategic use of air power in this last war? (Note the restricted field covered; consideration of the tactical use of air power in support of ground forces would require additional space beyond the scope of the present article.)

1. One lesson is that the time we were given to make our preparations was an absolutely essential factor in our final success. We had warning in 1939, and by 1941 had made notable progress. Following Pearl Harbor, with the United States actually at war, we had two and a half years more to build the striking force necessary to fulfill the strategic concept. The total time allowed us to prepare for the final all-out assault was four and a half years. It is unthinkable that we should ever again be granted such grace.

Under the A.A.F. expansion program after Pearl Harbor, the total personnel, the number of combat groups and the number of aircraft mounted steadily. On the other hand, the tonnage of bombs dropped in a month did not begin to rise significantly until early in 1944. It reached a peak around D-D ay, only to slacken off during the winter fogs of 1944-45, before attaining the all-time high prior to V-E Day. The gap between expansion in planes and personnel and the actual dropping of bombs tells the story of preparation for battle, of training, of technical supply, of adaptation and modification, of experimentation, of winning control of the air. It represents the time lag between the formation of tactical units and their conversion into striking power over the targets.

Had our peacetime air force been maintained during the 1930's at the level it attained even as early as the date of Pearl Harbor, and had it in consequence been prepared to act in the first year of war on the level it attained in mid-1942, then the tremendous and costly effort of the next two and a half years would have been enormously lessened. We would have struck at the heart of the enemy much earlier. It is even conceivable that the fact of an American air force in being, with full potential in 1939, might have prevented the outbreak of war.

In the next war, should there ever be one, four and a half years will not be allowed us in which to build up an air force, insured by the resistance of our Allies to common enemies. America will be Target Number 1; we will stand or fall with the air force available in the first crucial moment.

2. Air power in this war developed a strategy and tactic of its own, peculiar to the third dimension. It achieved the principle of mass, in the highest degree ever known, by its capacity to concentrate all its available units of striking power from widely distributed bases over one point — the enemy's heart. Any other force, operating in two dimensions, must strike at the periphery, the traditional line of war, and can reach the enemy's heart only after successful field campaigns. Air power at full potential overcomes the advantage of interior lines which centrally located countries previously enjoyed. It is not committed to battle, but returns to its base in preparation for a renewal of the assault. No other instrument of war has equivalent characteristics.

3. The first and absolute requirement of strategic air power in this war was control of the air in order to carry out sustained operations without prohibitive losses. The strategic offensive would not have been possible without the long-range fighter escort.

4. We profited by the mistakes of our enemies. The Germans were land-minded. In planning their aggression they did not allot their air force an independent mission of strategic offensive. Consequently they failed to meet their one historic opportunity to win decisively and quickly in 1940. Possibly their military leaders were fatally handicapped by the Nazi dictatorship. At any rate, they never recovered the advantage of air superiority in numbers over Britain, which later was to become the American base. They discovered too late the fatality of their lack of heavy bombers. They had been diverting plant capacity from making fighters to making V-1's and V-2's. But these arrived too late to affect the course of the war. Had they used the V-1 against shipping in the British ports prior to D-Day the invasion might perforce have been postponed for another year. After our inspection of their underground installations, we realized that their manufacture of jet fighters, and even jet bombers, could have reached dangerous proportions in another six months. These had been assigned first priority on the dwindling German oil supply. Given the super-speed of the jet-fighters, and given a sufficient supply of them (planned production: 1,200 per month), the Germans might have regained control of the air over Germany while we were waiting for our own jet production to catch up. In that contingency anything might have happened. Certainly, the end of the war would have been delayed.

To rely on the probability of similar mistakes by our unknown enemies of the future would be folly. The circumstances of timing, peculiar to this last war, and which worked out to our advantage, will not be repeated. This must not be forgotten.

5. Strategic Air Power could not have won this war alone, without the surface forces. The circumstances of timing did not permit. The full potential of sufficient striking power was attained only in the winter of 1943-44. By 1944 much of German war industry was going underground. Further, the invasion by land was necessary in order to force the diversion of German manpower from production, and even from manning the Luftwaffe. Thus, this war was won by the coordination of land sea and air forces, each of the Allies contributing its essential share to the victory. Air power, however, was the spark to success in Europe. And it is interesting to note that Japan was reduced by air power, operating from bases captured by the coordination of land, sea and air forces, and that she surrendered without the expected invasion becoming necessary.

Another war, however distant in the future, would probably be decided by some form of air power before the surface forces were able to make contact with the enemy in major battles. That is the supreme military lesson of our period in history.

NOTES

1 *Editor's Note [to Original Article]*: General Spaatz, then a Lieutenant-Colonel, was air observer, attached to the American Embassy in London, from May to September 1940. His official report that the Blitz would fail through German misuse of air power was one of the influential predictions of the war.

ABOUT THE AUTHOR

General Carl A. "Tooey" Spaatz was the top operational airman in charge of strategic bombing during WWII. After the war, he became the first Chief of Staff of the newly-created U.S. Air Force, and later served as CAP's first Chairman of the National Board.

From: Gen Carl Spaatz, "Strategic Air Power: Fulfillment of a Concept," *Foreign Affairs* 24 (1945/1946): 385-396. Used with permission.

14.2 Warden and the Air Corps Tactical School:

What Goes Around Comes Around

By Major Howard D. Belote, USAF

OBJECTIVES:

5. List the two questions of airpower theory that the Air Corps Tactical School sought to answer.
6. Define the operational context in which the ACTS theorists developed their airpower theories.
7. Define the operational context in which Col John Warden developed his airpower theories.
8. Describe, in your own terms, Kenneth Walker's "inviolable principle" for bombers.
9. Describe the major thematic differences between ACTS theory and Col Warden's airpower theory.
10. Name the three pathologies of airpower that affect both ACTS and Col Warden's contributions.

What has been will be again, what has been done will be done again; there is nothing new under the sun.
--Ecclesiastes 1:9

Between 1926 and 1940, officers at the Air Corps Tactical School (ACTS) created the theory and doctrine which would undergird the air strategies practiced in World War II. The "Bomber Mafia," which included Robert Olds, Kenneth Walker, Donald Wilson, Harold Lee George, Odas Moon, Robert Webster, Haywood Hansell, Laurence Kuter, and Muir S. Fairchild, sought to answer two basic questions of airpower theory. In the words of Lt Col Peter Faber, they asked, "What are the vital elements of an enemy nation's power and how can airpower sufficiently endanger them to change an opponent's behavior?"[1] To answer those questions, ACTS theorists portrayed nation-states as interconnected economic systems containing "critical points whose destruction will break down these systems" and posited that high-altitude precision bombing could effect destruction sufficient to achieve strategic objectives.[2]

Similarly, in the late 1980s, Col John A. Warden III developed the theoretical basis for the successful air strategy used in the Gulf War. Before the war, he wrote *The Air Campaign: Planning for Combat*, a balanced study of why and how to achieve air superiority. After becoming director of Checkmate, a Pentagon air strategy think tank, Warden focused on the strategic use of airpower. He created his "five rings" model and based Instant Thunder, Desert Storm's air operations plan, on it. Warden subsequently promulgated his ideas in essays such as "Air Theory for the Twenty-first Century" and "The Enemy as a System,"[3] which, like ACTS theory, depict strategic entities as definable systems with centers of gravity whose destruction can influence the system as a whole.

As examples of war-tested, uniquely American airpower theory, ACTS and Warden merit special examination. Interestingly, despite the 50 years separating their development, the theories have much in common in context and content. To demonstrate these similarities, this article compares and contrasts the history, central ideas, and assumptions of the theories. It then highlights their common strengths and weaknesses. Finally, those parallels are used to suggest lessons for twenty-first-century airpower thought.

BACKGROUND OF THE THEORIES

Historically, the two theories developed in similar contexts. As Faber notes, the ACTS theorists wrote to create a central role and mission for the fledgling Air Corps. Rapid demobilization after World War I had left the Air Service "chaotic, disorganized, [and] tangled," lacking both the equipment needed for training and "coherent theory, strategy, and doctrine upon which airmen could base the future development of American airpower."[4] Without such a working theory, airpower was likely to remain subordinate to Army traditionalists, who considered airplanes as a tool of the corps commander. Under Army control, airpower would be used primarily for observation and artillery spotting – certainly not for the strategic bombing concepts promoted by radicals like Billy Mitchell. Facing that threat, ACTS theorists posited a decisive strategic role for the precision bomber.

Similarly, John Warden wrote to fill a void in airpower discourse and to counter a trend of increasing subordination to the Army. Following the development of the atomic bomb, airmen left theory to civilians like Thomas Schelling and Bernard Brodie and tended to concentrate on technological issues The airmen appeared content with Brodie's observation that nuclear weapons made Giulio Douhet relevant, and they sought new and better ways of delivering atomic devastation to the enemy. However, when war experience in Korea and Vietnam proved that strategic bombing was insufficient, the focus gradually shifted from strategic to tactical airpower.

Faced by the Soviet threat during the 1970s and 1980s, American air leaders let the Army take the lead in developing doctrine. The result was the doctrine of AirLand Battle, and the Air Force accepted a supporting role. In *The Generals' War: The Inside Story of the Conflict in the Gulf*, Michael R. Gordon and Bernard E. Trainor note that in 1990 the commander to Tactical Air Command, Gen Robert D. Russ, and Lt Gen Jimmie Adams, Air Force deputy chief of staff for plans and operations, "believed that the Air Force's main role was to support the Army."[5] Warden, however, found both the old nuclear doctrine and the new supporting, attrition-based scheme "too limiting" and set out to prove that airpower, precisely directed against centers of gravity, could coerce political concessions from an enemy. In suggesting that airpower could dominate a conflict, Warden received the same cold shoulder the ACTS theorists had gotten 60 years earlier. His boss, General Adams, let Warden know that "his theorizing was radical."[6]

Interestingly, these contextual similarities – filling a theoretical gap while trying to avoid subordination to ground forces – gave rise to similar theories. Both ACTS and Warden used metaphors to describe, in Faber's words, "the vital elements of an enemy nation's power." Both theories focused on the enemy's will and capability to fight and portrayed states as closed systems that can be disrupted or paralyzed by destroying key targets. Finally, both theories prescribed courses of action based on similar assumptions. Examination of the central propositions of these theories will show that, despite some differences, the "industrial web" and the "five rings" are kindred spirits.

CORE PROPOSITIONS

Central to the ACTS theory was the notion that economic destruction would lead to social collapse and enemy capitulation. ACTS theorists described enemy systems variously as a "precision instrument," "wispy spider's web," or "tottering house of cards."[7] Haywood S. Hansell fleshed out the argument as follows:

1. Modern great powers rely on major industrial and economic systems for production of weapons and supplies for their armed forces, and for manufacture of products and provision of services to sustain life in a highly industrialized society. Disruption or paralysis of these systems undermines both the enemy's *capability* and *will* to fight [emphasis in original].

2. Such major systems contain critical points whose destruction will break down these systems, and bombs can be delivered with adequate accuracy to do this.

3. Massed air strike forces can penetrate air defenses without unacceptable losses and destroy selected targets.

4. Proper selection of vital targets in the industrial/economic/social structure of a modern industrialized nation, and their subsequent destruction by air attack, can lead to fatal weakening of an industrialized enemy nation and to victory through air power.[8]

The "fatal weakening" resulting from these attacks against enemy capability and will was so important that it precluded using bombers in any other role. Kenneth Walker set forth an "inviolable principle": The bomber must only fly against "vital material targets" deep in the enemy heartland and never in Army support.[9] To do otherwise would be to squander the bomber's power.

To focus the bomber's power appropriately, the ACTS theorists sought to identify those critical points that would bring down the enemy system. Harold Lee George first suggested that by attacking "rail lines, refineries, electric power systems, and (as a last resort) water supply systems...an invader would quickly and efficiently destroy the people's will to resist."[10] Robert Webster and Muir Fairchild refined George's list of "will" targets. They focused specifically on "national *organic systems* on which many factories and numerous people depended" [emphasis in original].[11] According to Hansell, organic systems included production and distribution of electricity, fuel, food, and steel; transportation networks; and certain specialized factories, especially those producing electrical generators, transformers, and motors.[12] Despite a lack of economic intelligence – theorists identified the foregoing systems by studying the United States – ACTS predicted victory for those who followed the "industrial web" prescriptions.

Roughly half a century later, John Warden applied a new metaphor to the ACTS vision of the enemy as a system. Fortified by his knowledge of military theory – specifically, that of J. F. C. Fuller – and modern communications technology, Warden followed a traditional practice and likened the enemy system to the human body. Rather than an amorphous "web" or "house of cards," Warden described an enemy (indeed, every life-based system) as an entity with a brain, a requirement for "organic essentials," a skeletal-muscular infrastructure, a population of cells, and a self-protection mechanism. He arranged these components into the now-familiar model of five concentric rings, with each ring dependent on the ones inside it. Warden's major additional to ACTS theory – the brain, or leadership ring – controlled the entire system. If the center ring could be killed (Fuller's "shot through the head"), or isolated by severing communications links, the entire system would crumble.[13]

Just like the ACTS theorists, Warden focused on the enemy's will and capability to fight. "It is imperative," he argued, "to remember that all actions are aimed against the enemy system as a whole." Furthermore, "when the command element cannot be threatened directly, the task becomes one of applying sufficient indirect pressure so that the command element rationally concludes that concessions are appropriate, realizes that further action is impossible, or is physically deprived of the ability to...continue combat."[14] If unable, then, to attach the center leadership ring directly, Warden recommended attacks on organic essentials such as power production and petroleum – precisely the targets identified by ACTS. He proposed that damage to organic essentials could lead to "collapse of the system" or "internal political or economic repercussions that are too costly to bear"[15] – in other words, to the "fatal weakening" suggested by ACTS. Finally, just as the ACTS theorists refused to squander bombing on Army support operations, Warden emphasized that "engagement of the enemy military...should be avoided under most circumstances." Fighting an enemy's military "*is at best a means to an end and at worst a total waste of time and energy*" [emphasis in original].[16]

In essence, Warden just updated ACTS theory. The major thematic difference between the theories is the addition of a new "vital center" – the leadership ring – and two new destructive mechanisms to influence that center of gravity: *decapitation and parallel war*. Nuclear strategists coined the first term to describe the killing or isolation of enemy leaders; Warden created the second to describe the overwhelming-force strategy to use when the leaders were unreachable. A "death of 1,000 cuts" would suffice to collapse an enemy system whose center ring was protected, just as ACTS proposed to disrupt the industrial web. Technology improved the execution of the strategy, however, allowing airmen to inflict those cuts nearly simultaneously. Warden noted that Desert Storm air forces "struck three times as many targets in Iraq in the first 24 hours as Eighth Air Force hit in Germany in all of 1943."[17]

UNDERLYING ASSUMPTIONS

Given the similarities in context and content that connect these bodies or airpower thought, it should not be surprising to discover that they rest on similar assumptions. Most importantly, they presuppose a rational actor, or, to use Graham Allison's term, *Model I enemy*. Warden proposed that "enemies, whether they be states, criminal organizations, or individuals all do the same thing: they almost always act or don't act based on some kind of cost-benefit ratio."[18] Faber made the same observation about ACTS, whose theorists overlooked the fact that an enemy might operate based on "potentially obscure organizational, bureaucratic, or emotional" Model II/III factors.[19] Faber also pointed out that ACTS theory rested on a "mid-Victorian faith in technology" and "wrongly assumed that revolutionary bomber-related technologies would produce almost 'frictionless' wars."[20] Warden echoed this faith, consigning friction to the Napoleonic era. In Warden's combat equation, modern airmen could ignore morale (and friction, a morale-related factor) because physical factors x morale = outcome. When physical factors approach zero due to technologically superior attacks, output of the enemy war machine will be zero, regardless of morale factors – and friction is therefore irrelevant.[21]

Clearly, these assumptions lead to problems. Due to its simplicity, a rational-actor model cannot adequately describe or predict the behavior of many state and non-state actors. Faber, for example, asks, "Is it not possible... that a state might continue to struggle – at higher costs – to demonstrate its resolve in future contingencies?"[22] If a strategist cannot determine how an opponent will react to pressure – if the Model I analysis is faulty – then he cannot effectively target the opponent's will or force him to change his mind à la Warden and ACTS. A belief in frictionless war seems fraught with peril, as well. Gordon and Trainor devote a full chapter to describing numerous instances of friction in the Gulf War; Lt Col Barry D. Watts uses an entire book to show how twentieth-century warfare is characterized by friction. "The very structure of human cognition," he concludes, "argues that friction will continue to be the fundamental atmosphere of war."[23] These flawed underlying assumptions cast doubt on the validity of both theories and suggest additional questions. Do the ACTS and Warden theories share other flaws? If they do, are they relevant to airpower strategists in the coming years?

HOLES IN THE LOGIC

The theories do, in fact, contain additional related flaws that highlight lessons for future strategists. Faber characterizes these flaws as the "three pathologies" of airpower theory. One of the pathologies is an overreliance on metaphor in place of logical argumentation.[24] ACTS theorists and Warden provided little evidence to support their "web" and "body" analogies. Warden merely rearranged a tabular presentation of system components into rings and claimed – without empirical data – that the diagram proved "several key insights," namely that the rings were interdependent, the center was most important, that the military was merely a shield for the others, and effectiveness lay in working inside-out vice outside-in.[25] Warden also failed to provide proof that a nation-state, like a body, could be killed through decapitation. Similarly, the ACTS theorists described an economic "house of cards" using a sample size of one – the American economy of the 1930s.

Critiquing Warden, Dr. Lewis Ware notes that such unsupported metaphors are inadequate as analytical instruments. Their "arguments rest on principled belief rather than on reason, and principled belief – however powerful or well intended – is by definition not susceptible to rational explanation."[26] Faber points out that, unlike a human body, a society can substitute for lost vital organs; he further notes that metaphor-based theories have led to faulty employment of airpower in war because they fail to see that conflict is nonlinear and interactive.[27] The message for strategists is clear: Examine theoretical metaphors carefully. Ensure that verifiable cause-and-effect relationships exist between the parts of a metaphor that provide its explanatory power, especially if the metaphor is used to plan an air strategy. Finally, remember that enemies react. Decision makers should not expect an Iraqi-style rollover.

ACTS and Warden share Faber's second "pathology" as well: They both "made a fetish of quantification and prediction in war."[28] As Faber notes, the ACTS instructors who wrote Air War Plans Division – Plan I calculated precisely how to defeat Germany: 6,960 bombers attacking 154 target sets would produce victory in six months. Likewise, Warden claimed that "with precision weapons, even logistics become simple...[S]ince we know that all countries look about the same at the strategic and operational levels, we can forecast in advance how many precision weapons will be needed to defeat an enemy."[29]

Political scientist Robert Pape has highlighted the problem with such quantification. Strategists who rely on predictions like the forecasts cited above confuse combat effectiveness with strategic effectiveness. Operators should be concerned with the first, which concerns target destructions, while strategists and commanders must focus on the second and ask whether or not said destruction achieves political goals. Strategists cannot allow a quantitative focus to obscure their understanding of the human interaction that constitutes both war and politics. Despite Warden's claims to the contrary, technology has not invalidated Clausewitz; war is still unpredictable.

The unwavering devotion with which ACTS theorists and Warden clung to the aforementioned "pathologies" highlights their susceptibility to Faber's final pathology. Faber notes that "air theorists sought to develop hoary maxims that would apply to all wars, regardless of time and circumstance. The ACTS 'Bomber Mafia,' for example, adopted 'a Jominian, mechanistic view of war – a view of war as a mathematical equation whose variables can be selectively manipulated to achieve success.'"[30] Warden's previously cited "outcome" equation and his claim that the five rings are "general concepts not dependent on a specific enemy" suggest that he also believed in a universally applicable strategic formula. Both theories, however, ignore the role of historical, cultural, and moral context, and that limits their universality.[31] More importantly, their claims of universality have led to widespread skepticism.

Arguably, that skepticism underlies the current battles over airpower's role in joint doctrine. Gen Ronald R. Fogleman has said that, due to the claims of airpower visionaries, "we found ourselves in a position where there were a lot of unfulfilled promises and false expectations relative to what airpower could and could not do." He further admonished airmen not "to let our enthusiasm for our primary mediums of operations blind us to the advantages that can be gained by using airpower in support of land and naval component objectives." [32] He suggested that airmen are partly to blame for current interservice battles. In other words, the adherence of air theorists to "hoary maxims" has hampered the development of joint doctrine. Future air strategists can alleviate that problem by claiming less universality for airpower ideas.

THE BOTTOM LINE

Do these pathologies inherent in the ideas of ACTS and Warden invalidate the theories? No. Warden critic Lewis Ware admits that Warden's "reductionism has immense practical value for the successful prosecution of an air action."[33] Col Richard Szafranski is more blunt: "Purism matters less to action-oriented people than the verifiable consequences of action...Try as critics might, they cannot eradicate the objective reality of the Desert Storm air battles. They worked."[34] Similiarly, after a long trial and midcourse adjustments, ACTS theory succeeded. By late 1944, attacks on fuel production and transportation nearly prevented German forces from flying or driving at all. Szafranksi's critique of Warden applies equally to ACTS: Each "dares to offer us a map for air warfare. Its imperfections does not erase its utility... [If] 'bold ideas, unjustified anticipations, and speculative thought are our only means...we must hazard them to win our prize.'"[35] ACTS theorists and John Warden provided frameworks for winning air campaigns.[36] Despite their common flaw, the theories provide valuable understanding of air warfare and starting points for further theoretical development.

In the 1920s and 1930s, ACTS theorists proposed an answer to the "two basic questions of airpower theory": (1) What are the vital elements of an adversary's power? (2) How can airpower influence them? Writing to prevent a subordinate role for airpower, the ACTS instructors suggested that nations could be coerced or destroyed by precision bombing of their "industrial web." In the 1980s and 1990s, John Warden updated ACTS theory. He wrote in a similar context, added a leadership ring to the economic target list, and echoed ACTS's claims about precision. Both theories lay on questionable assumptions about enemy rationality and technology's ability to overcome friction, and both fell prey to Faber's "pathologies" of airpower theory – overreliance on metaphor and quantification, and a Jominian claim to universality. In the final analysis, however, both worked. Air strategists can, therefore, learn much from the shortcomings and strengths of the airpower theories of the Air Corps Tactical School and Col John Warden – and future theorists have therein a ready-made, battle-tested foundation for shaping the aerospace power of the next century.

NOTES

1. Lt Col Peter Faber, "Competing Theories of Airpower: A Language for Analysis," paper presented at the Air and Space Power Doctrine Symposium, Maxwell AFB, Ala., 30 April 1996. Available on-line from http://www.airpower.maxwell.af.mil/airchronicles/presentation/faber.html

2. Lt Col Peter Faber, "Interwar US Army Aviation and the Air Corps Tactical School: Incubators of American Airpower," in Col Phillip S. Meilinger, ed., Paths of Heaven: The Evolution of Airpower Theory (Maxwell AFB, Ala.: Air University Press, 1997), 217.

3. Col John A. Warden III, "Air Theory for the Twenty-first Century," in Kari P. Magyar, ed., Challenge and Response (Maxwell AFB, Ala.: Air University Press, 1994), 311-32; and "The Enemy as a System," Airpower Journal 9, no. 2 (Spring 1995): 40-55.

4. Faber, "Interwar US Army Aviation," 185.

5. Michael R. Gordon and Bernard E. Trainor, The Generals' War: The Inside Story of the Conflict in the Gulf (Boston: Little, Brown, and Co., 1995), 79.

6. Ibid.

7. Faber, "Competing Theories," 1-2.

8. Quoted in Faber, "Interwar US Army Aviation," 217.

9. Ibid., 219.

10. Ibid., 194.

11. Hansell, quoted in ibid., 219.

12. Ibid.

13. Warden, "Air Theory," 311-32; and "The Enemy as a System," 40-55. For Fuller's influence on Warden, see Lt Col David S. Fadok, "John Boyd and John Warden: Airpower's Quest for Strategic Paralysis," in Meilinger, 361.

14. Warden, "The Enemy as a System," 49.

15. Ibid.

16. Warden, "Air Theory," 317-18.

17. Ibid., 324. Of course, there are other differences between the theories; for example, ACTS assumed total war with maximum destruction, while Warden foresaw limited war with minimum collateral damage – another update which reflected better technology and, perhaps, the "CNN factor." These differences are peripheral, however.

18. Ibid., 314.

19. Faber, "Interwar US Army Aviation," 221. See also Graham Allison, Essence of Decision: Explaining the Cuban Missile Crisis (Boston: Little, Brown and Co., 1971) for discussion of Models I, II, and III (Rational Actor, Organizational Process, and Bureaucratic Politics) analyses.

20. Ibid., 220. ACTS instructors did, in fact, include "fudge factors" in their calculations, but they turned out to be far too small.

21. Warden, "The Enemy as a System," 42-43.

22. Faber, "interwar US Army Aviation," 221.

23. Lt Col Barry D. Watts, The Foundations of U.S. Air Doctrine: The Problem of Friction in War (Maxwell AFB, Ala.: Air University Press, 1984), 93.

24. Faber, "Competing Theories," 1-2.

25. Warden, "Air Theory," 315-17

26. Dr. Lewis Ware, "Ware on Warden: Some Observations of the Enemy as a System," Airpower Journal 9, no. 4 (Winger 1995), 92.

27. Faber, "Competing Theories," 2.

28. Ibid., 1.

29. Warden, "Air Theory," 327-28.

30. Faber, "Competing Theories," 1; internal quotes from Col Thomas A. Fabyanic, "War Doctrine and the Air War College – Some Implications for the U.S. Air Force," Air University Review 37, no. 2 (January-February 1986).

31. See Faber, "Competing Theories," 1; and Ware, 91, on the lack of contextual understanding of ACTS and Warden, respectively.

32. Gen Ronald R. Fogleman, "Aerospace Doctrine – More Than Just a Theory," keynote address to the Air and Space Power Symposium, Maxwell Air Force Base, Alabama, 30 April 1996. Available on-line from http://ww.airpower.maxwell.af.mil/chronicles/presentation/doctrel.thml.

33. Ware, 89.

34. Col Ricahrd Szafranski, "The Problem with Bees and Bombs," Airpower Journal 9, no. 4 (Winger 1995): 96.

35. Ibid., 97. Szafranski attributes the internal quotation to Karl R. Popper as cited in Timothy Ferris, ed., The World Treasury of Physics, Astronomy, and Mathematics (Boston: Little, Brown and Company, 1991), 799.

36. It is true that the Gulf War theater planners in the "Black Hole" made substantial additions to Warden's original scheme; they certainly deserve credit for their contributions to airpower theory and coalition victory.

ABOUT THE AUTHOR

Howard D. Belote wrote this article while a student at the prestigious School of Advanced Airpower Studies at Maxwell AFB, Ala. Before retiring in 2010, Col Belote completed command assignments at the squadron, group, and wing levels.

From: Maj Howard D. Belote, "Warden and the Air Corps Tactical School: What Goes Around Comes Around," *Airpower Journal* (Fall 1999): 39-47. Used with permission.

14.3 Cyberspace: The New Air and Space?

By Lt Col David A. Umphress, USAFR

OBJECTIVES:

11. Define the term "cyberspace."
12. List two benefits of operating in a rich cyberspace environment.
13. Describe the four fundamental principles of cyberspace.

The mission of the United States Air Force is to deliver sovereign options for the defense of the United States of America and its global interests—to fly and fight in Air, Space, and Cyberspace.
—USAF Mission Statement

In late 2005, the Air Force altered its mission statement. As before, the service flies and fights in air and space, but now it also flies and fights in *cyberspace*. We have long recognized that information serves as a center of gravity for the military. Although military operations may involve aircraft, guns, tanks, ships, and people, information is the "glue" that tells each aircraft what sortie to fly, each tank where to go, and each ship where to sail. The revised mission statement represents a bold move if for no other reason than the fact that its explicit mention of cyberspace brings to the forefront the role played by information and information technology in the modern Air Force. Indeed, the statement elevates the notion of cyberspace and its attendant infrastructure to the level of importance occupied by air and space. Whereas, formerly, the Air Force perceived itself as carrying out kinetic operations, the latest version of its mission statement places the service squarely in the nonkinetic arena as well.

We have an intuitive sense of how the Air Force operates in air and space since both are physical in nature. Less clear is the relationship between the Air Force and cyberspace. What is cyberspace? Why is it important? What are the rules under which it operates?

CYBERSPACE DEFINED

In the early 1980s, writer William Gibson coined the term cyberspace to describe a fictionalized computer network containing vast amounts of information that could be tapped for wealth and power.[1] In his cyberspace, the physical world and the digital world become blurred to the point that human users perceive computer-generated experiences that have no real existence, and sentient digital beings affect the physical world. Although Gibson's depictions of computer-simulated reality, cybernetically enhanced humans, and artificially intelligent entities remain in the realm of science fiction, the concepts of "exploring" vast amounts of data and "visiting" remote computers do not. Moreover, the premise that computer networks contain information that people can exploit—for good and ill—is very real.

We need a physical infrastructure of computers and communication lines to implement cyberspace. In other words, cyberspace runs "on" computers. However, what resides "inside" computers provides the greatest leverage: we measure the true value of cyberspace in terms of the information contained within that infrastructure. The crucial characteristics of cyberspace include the fact that (1) information exists in electronic format, and (2) computers can manipulate (store, search, index, process, etc.) that information.

Cyberspace has thus become a metaphor for the digital society made possible through computers and computer networks. When referred to abstractly, it connotes the sum total of information available electronically, the exchange of that information, and the communities which emerge from the use of that information. When used in reference to a particular military operation, it signifies the information available to a specific audience.

Cyberspace need not be publicly accessible although the public does have access to the predominant implementation of cyberspace— the Internet. Military units can operate private networks that constitute their own limited versions of cyberspace. In fact, many disconnected "cyberspaces" can exist simultaneously, each servicing its own community of users.

WHY CYBERSPACE IS RELEVANT

Marshall McLuhan's aphorism "the medium is the message" characterizes our expectations of cyberspace. He points out that "societies have always been shaped more by the nature of the media by which humans communicate than by the content of the communication."[2] Since computers and electronic communication networks encourage the rapid and widespread exchange of information, it naturally follows that they would also influence military operations.

It is interesting to observe the evolution of the medium-is-the-message effect on the Air Force's perception of cyberspace. Initially, government policies equated cyberspace with the communication hardware comprising computer networks, concentrating on hardening to protect against infiltration. Later policies envisioned cyberspace not only as networks but also as the data transmitted across them, which led to a focus on data integrity. The change in the Air Force's mission statement to include cyberspace implies that we now perceive it as *content*—something more than hardware and data.

The electronic encoding of information in cyberspace, rather than on physical media, permits wider interchange of those data. This is the foundation of an information-driven society proposed over the last 30 years by so-called new-age pundits such as McLuan, John Naisbitt, Alvin Toffler, and Don Tapscott, to name a few.[3] The premise of the information society is that information itself has economic value, with a corollary which holds that information has operational value to the military. The more efficiently and effectively we manage information, the more benefit we derive from it.

The military has recognized this idea by declaring "information superiority" as one of its core values.[4] It has moved to organize and equip itself so as to improve the management of information. The specific organizational approaches have various names—net-centric, knowledge management, battlespace, infosphere, and so forth—but the general concept remains the same: create a rich cyberspace (with tools, sensor-provided data, quality of information, etc.) in which to make decisions.[5]

Ideally, two primary benefits become evident from operating in such an information-driven environment. First, the organization can be decentralized as much as is feasible within a military context. Everyone operates within cyberspace and has access to the appropriate information needed to make decisions. We no longer have to make decisions at the point in the organization determined by the nexus of suitable information, but at the point most affected by the decision. Second, the organization can function as a coalition of semi-independent agents whose environment drives their operations.

For every benefit, however, a host of side effects exists. Technology that relies on information encoded in electronic format remains central to supporting information superiority. That technology does not exist in any integrated fashion today. We carry out information-related functions with a patchwork collection of software and hardware tools. We also struggle with a number of questions: How do we manage massive amounts of information? How do we prevent the mining of large amounts of unclassified data for classified information? How do we "compartmentalize" cyberspace so that the right information gets to the right decision makers? What information can we transmit over unclassified civilian networks versus tightly controlled, classified military networks? How do we integrate information coming through official military networks with information coming from "back-channel" sources? How computer savvy do users of cyberspace have to be? What mechanisms are in place to detect information tampering?

FUNDAMENTAL PRINCIPLES OF CYBERSPACE

The Air Force's announcement of its revised mission statement prompted a considerable amount of discussion regarding the precise definition of *cyberspace* and the way it relates to air and space. In the midst of this discussion were debates about what constitutes the bounds of cyberspace, whether it can function as a medium for weapon delivery, how the Air Force *flies* through cyberspace, and the like. That this discussion arose demonstrates that the concept of cyberspace is very much open to debate. As with the proverbial blind men giving their interpretations of the elephant, we have a number of ways of looking at cyberspace, depending on our perspective. Regardless of how we ultimately view cyberspace, though, we must recognize that it operates under some fundamental principles.

Information Is the Coin of the Realm in Cyberspace

Since cyberspace deals with information, the latter naturally determines the "economy" of the particular cyberspace in which it resides. In other words, we can think of information as having "value," which depends on its inherent usefulness as a stand-alone piece of information as well as the way it relates to other information, both within cyberspace and without. Changes in the availability or usefulness of the information alter its value.

For example, content on an intranet page may gain in value if it leads to other information of equal or greater value. Similarly, it may lose value if it is duplicated or contradicted somewhere else. In the absence of relationships with other information, the value of information in cyberspace generally decreases over time because it has a greater chance of having been put to some use.

We need not restrict the notion of value to factual information. There is no guarantee regarding the accuracy or truthfulness of information in cyberspace. Consequently, *disinformation* intended to disguise the worth of legitimate information has value.

We may not explicitly know the value of a particular piece of information in cyberspace. Certainly, if it has a security classification, we understand the inherent risk if that information is compromised. We thus attach an arbitrarily high value to such information. However, it is computationally infeasible to compare one piece of information to all other combinations of pieces of information within cyberspace in order to determine value. We cannot know, a priori, when we can combine a particular piece of information, classified or not, with another piece of information to form intelligence higher in value than the individual pieces separately. To complicate things further, hardware and software appliances that "sniff" networks and intercept data transmissions often prevent us from determining if someone has obtained a piece of information illicitly, thereby unknowingly altering its value. Encryption and other information-assurance measures mitigate such occurrences to a great extent but don't prevent them.

Paradoxically—at least in terms of economic theory—the ever-increasing supply of information available within cyberspace does not decrease the value of information. Instead, its value increases due to the scarcity of time and resources required to find useful information from the overall supply. This phenomenon has given rise to "technopower," the concept that power and control are in the hands of people able to use cyberspace technology effectively to obtain high-value information.[6]

Cyberspace Shapes Authority

Although information itself defines value in cyberspace, access to that information determines power and, consequently, shapes authority. Economists portray information as falling into one of three categories: free, commercial, and strategic.[7] Free information is available to whoever seeks it; commercial information to people willing to pay for it; and strategic information only to those specially entrusted to have it. Outside the context of cyberspace, strategic information has the greatest persuasive value because its restricted availability can serve as a source of influence and power over those who don't have it. Holders of strategic information serve as gatekeepers, doling it out as necessary for their own purposes.

The emergence of cyberspace has altered this balance of power, providing a mechanism for disseminating information widely and freely. Previously, we funneled and filtered valuable information through gatekeepers; now, however, we can bypass them altogether, thus permitting peer-to-peer communication of information. Given this model, strategic information will undergo almost instantaneous devaluation if we put it into cyberspace without providing some sort of protection because it becomes available to all users of that cyberspace. Further, making information freely available means it becomes more accessible and has the potential to reach a larger audience.

This scenario has had societal effects, the most profound of which are virtual communities. Whether implemented as a private network supporting military operations or as a public Internet, cyberspace connects people. Users of a military cyberspace are fairly homogeneous; their goals address a specific military operation. As the user base of cyberspace becomes larger and more public, not only do user goals diversify, but also communities form within cyberspace.

Take the Internet, for example. With an estimated audience of 1.8 billion users across 225 countries, it has transformed the globe into a virtual village.[8] People can communicate with each other regardless of physical location. In so doing, they are able to form and join social networks consisting of individuals with similar interests. The popularity of Web-based social networking tools such as Facebook (7 million users), Xanga (40 million), MySpace (108 million), and Hi5 (40 million) demonstrates the potential of cyberspace to bring people together.[9] [*Obviously these figures are dated.*]

This ability is not lost on nonstate actors, who use the Internet as a meeting place, recruiting tool, and conduit for propaganda. For example, Hezbollah has leveraged cyberspace technology quite effectively, sponsoring a number of Arabic and English Web sites that describe world events from a Hezbollah perspective. Its graphic pictures, video clips, and news articles of the Israel-Lebanon conflict in July 2006 are clearly designed to portray Israel as a terrorist puppet of the United States.[10] Realizing that many Israelis visit these sites, Hezbollah uses them to demoralize this Israeli audience while simultaneously boasting of its victories to the Arab audience.[11]

Cyberspace Operates under Nontraditional Physics

The juxtaposition of cyberspace with air and space in the Air Force's mission statement almost depicts cyberspace as a physical means for conducting operations. True, it is useful at some level of abstraction to conceptualize cyberspace as a medium. After all, cyberspace works through the medium of computers and networks. However, drawing too close an analogy between a physical entity (air and space) and a logical one (cyberspace) can be dangerous. Cyberspace operates on entirely different laws of physics than does physical space. For example, information doesn't weigh anything. It has no physical mass. It can instantaneously pop into—and out of—existence. It can be replicated without cost, accumulated without human intervention, and divorced from its physical location. Information does not, in itself, kill. It does so only when we use it to influence physical players in air and space. Because of the nonphysical nature of information, placing it in cyberspace gives it instant, global availability to all users of that cyberspace. We often cannot determine whether information we obtain from a source in cyberspace is original or has been copied from somewhere else within cyberspace.

Cyberspace—particularly the Internet—is a global phenomenon. Information that the United States does not wish to reveal may be available through sources located in countries outside its purview. We cannot necessarily control all information, nor can we necessarily remove a piece of information. We can only regulate information within our own span of control.

Cyberspace Brings the Front Line to the Front Door

Census and survey data indicate that 54 million households in the United States have at least one personal computer and that roughly two-thirds of Americans actively use the Internet in some fashion.[12] Fifty-seven million employed Americans—62 percent of the workforce— report using a computer at work, 98 percent of whom have access to electronic mail.[13] Of those, the majority reports trusting the content of electronic mail when it contains at least one item of personal information other than first name. We can reasonably assume that these statistics generally represent the Air Force workforce, given the 15 million personal computers in the Department of Defense's inventory, combined with the leadership's vision of a net-centric force.[14]

We can access public cyberspace literally from within our own homes or places of employment. For the first time in history, we have a vast amount of information at our fingertips. Also for the first time, we have the front line of a battle at our front door. Prior to cyberspace's rise in popularity, the main participants in military operations were soldiers physically engaged in conflict. News reports that portrayed the results of military action to civilians at home dealt with events happening outside the country's borders. With cyberspace within easy reach of ordinary citizens, those who wish to use it for ill gain have direct entrée into the home. This situation is particularly poignant since empirical studies have shown that computers, at home or otherwise, are probed for security vulnerabilities during the first 20 minutes of their connection to a public network.[15]

Contrary to the prevailing picture painted by the media, "war" in cyberspace will not likely manifest itself as an electronic Pearl Harbor, causing massive destruction. More probably, cyberwar will take the form of influence rather than lethality. Cyber warriors will not destroy infrastructure because that would be self-defeating, particularly within the United States. Instead, they will more likely obtain information they can use to manipulate happenings in the physical world to their advantage.

Those who choose to operate in cyberspace have a number of asymmetrical advantages. First, the "battlefield" is large and easy to hide in. Second, the effects of attacks are disproportionate to their costs. Using cyberspace is neither material- or capital-intensive. Individuals can access it with inexpensive computers, free software, and consumer-ready communication equipment. They can launch attacks from across the globe almost with impunity because of the difficulty of determining the exact origin of the attack or the identity of the attacker. Third, the one-sided nature of cyber attacks forces potential victims into assuming a defensive posture. The victim curtails his computer and communication services to within what his governance structure deems "acceptable," based on its perceptions of the prevailing dangers—real or not. In case of an attack, the victim probably will not launch an in-kind offensive action since, even if he can identify the attacker, he probably lacks the computer infrastructure to make a counterattack worthwhile.

CONCLUSION

Perhaps the greatest lesson we can derive from the Air Force's revised mission statement is that it warns all Airmen of the reality of cyberspace. The statement requires us to understand the implications of an information-reliant military. It also challenges us to look for ways to best use cyberspace and to understand that we can attain "throw weight" by finding new ways to make the best use of cyberspace technology.

B. H. Liddell Hart's admonition that a "strategist should think in terms of paralyzing, not killing" remains as relevant today as it ever was.[16] Although Liddell Hart spoke of paralyzing armies of people and the economies of states, his words nevertheless apply to the individual Airman. Never in history have so many people found themselves intimately tied to a weapon system—cyberspace—that is limited only by the human imagination.

NOTES

1. William Gibson, Neuromancer (New York: Ace Books, 1984).

2. Marshall McLuhan and Quentin Fiore, The Medium Is the Message (New York: Random House, 1967).

3. Ibid.; John Naisbitt, Megatrends: Ten New Directions Transforming Our Lives (New York: Warner Books, 1982); Alvin Toffler, The Third Wave (New York: Morrow, 1980); and Don Tapscott, The Digital Economy: Promise and Peril in the Age of Networked Intelligence (New York: McGraw-Hill, 1996), 6.

4. Joint Vision 2010 (Washington, DC: Joint Chiefs of Staff, 1996), 18.

5. John G. Grimes, "From the DoD CIO: The Net-Centric Information Enterprise," CrossTalk: The Journal of Defense Software Engineering 19, no. 7 (July 2006): 4; Managing Knowledge @ Work: An Overview of Knowledge Management (Washington, DC: Chief Information Officers Council, 2001), 7; Dr. David S. Alberts, Defensive Information Warfare (Washington, DC: National Defense University Press, August 1996), http://www.ndu.edu/inss/books/books%20 -%201996/Defense%20Information%20Warfare%20- %20Aug%2096/index.html; and Michael Vlahos, "Entering the Infosphere," Journal of International Affairs 51, no. 2 (1998): 497-525.

6. Tim Jordan, "Cyberpower: The Culture and Politics of Cyberspace," Internet Society, http://www.isoc.org/inet99/proceedings/3i/3i_1.htm.

7. Robert O. Keohane and Joseph S. Nye Jr., "Power and Interdependence in the Information Age," Foreign Affairs 77, no. 5 (September/October 1998): 89-92.

8. "Web Worldwide," ClickZ Stats, http://www.clickz.com/showPage.html?page=stats/web_worldwide.

9. Wikipedia: The Free Encyclopedia, s.v. "List of Social Networking Websites," http://en.wikipedia.org/wiki/List_of_social_networking_websites.

10. "Israeli Aggression on Lebanon," Moqawama.org, http://www.moqawama.org/aggression/eindex.php.

11. Maura Conway, "Cybercortical Warfare: The Case of Hizbollah.org" (paper prepared for presentation at the European Consortium for Political Research [ECPR] Joint Sessions of Workshops, Edinburgh, United Kingdom, 23 March-2 April 2003), http://www2.scedu.unibo.it/roversi/SocioNet/Conway.pdf.

12. Home Computers and Internet Use in the United States: August 2002 (Washington, DC: US Department of Commerce, US Census Bureau, September 2001), http://www.census.gov/prod/2001pubs/p23-207.pdf; and "U.S. Internet Adoption to Slow," ClickZ Stats, http://www.clickz.com/showPage.html?page=3587496#table1.

13. "Email at Work," Pew Internet and American Life Project, 8 December 2002, http://www.pewinternet.org/ reports/toc.asp?Report=79.

14. Army Field Manual 100-6, Information Operations, 27 August 1996, 14.

15. Lorraine Weaver, "They're Out to Get Us! The Cyber Threat to the Telecommuter, Small Office / Home Office (SOHO), and Home User" (presentation to the 14th Annual Systems and Software Technology Conference, Salt Lake City, UT, 1 May 2002), http://www.sstc-online .org/Proceedings/2002/SpkrPDFS/WedTracs/p1371.pdf.

16. B. H. Liddell Hart, Paris: Or the Future of War (New York: Garland Publishing, 1975), 40-41.

ABOUT THE AUTHOR

Lieutenant Colonel Umphress is an individual mobilization augmentee at the College of Aerospace Doctrine, Research and Education, Maxwell AFB, Alabama, and an associate professor in the Department of Computer Science and Software Engineering at Auburn University.

From: Lt Col David A. Umphress, "Cyberspace: The New Air and Space?," *Air & Space Power Journal* (Spring 2007): 50-55. Used with permission.

14.4 Basic Air Force Doctrine

AF Doctrine Document 1-1 (2011)

Selections from Air Force Doctrine Document 1, *Air Force Basic Doctrine, Organization, and Command,* Chapters 2 & 5

OBJECTIVES:

14. Define the term "air power."
15. Describe how air power has evolved from strategic (kinetic) bombing to encompass information, space, and cyberspace operations.
16. Identify key attributes of air power.
17. Define the term "airmindedness."
18. Recall some practical applications of airmindedness for Airmen.
19. List and briefly define he twelve Air Force core functions.

The other services have air arms—magnificent air arms—but their air arms must fit within their services, each with a fundamentally different focus. So those air arms, when in competition with the primary focus of their services, will often end up on the short end, where the priorities for resources may lead to shortfalls or decisions that are suboptimum. It is therefore important to understand that the core competencies of [airpower] are optional for the other services. They can elect to play or not play in that arena. But if the nation is to remain capable and competent in air and space, someone must pay attention across the whole spectrum; that is why there is a US Air Force.

— General Ronald R. Fogleman, USAF, retired

AIRPOWER

Airpower is the ability to project military power or influence through the control and exploitation of air, space, and cyberspace to achieve strategic, operational, or tactical objectives. The proper application of airpower requires a comprehensive doctrine of employment and an Airman's perspective. As the nation's most comprehensive provider of military airpower, the Air Force conducts continuous and concurrent air, space, and cyberspace operations. The air, space, and cyberspace capabilities of the other Services serve primarily to support their organic maneuver paradigms; the Air Force employs air, space, and cyberspace capabilities with a broader focus on theater-wide and national-level objectives. Through airpower, the Air Force provides the versatile, wide-ranging means towards achieving national objectives with the ability to deter and respond immediately to crises anywhere in the world.

Airpower exploits the third dimension of the operational environment; the electromagnetic spectrum; and time to leverage speed, range, flexibility, precision, tempo, and lethality to create effects from and within the air, space, and cyberspace domains. From this multi-dimensional perspective, Airmen can apply military power against an enemy's entire array of diplomatic, informational, military, and economic instruments of power, at long ranges and on short notice. Airpower can be applied across the strategic, operational, and tactical levels of war simultaneously, significantly increasing the options available to national leadership. Due to its range, speed, and flexibility, airpower can compress time, controlling the tempo of operations in our favor. Airpower should be employed with appropriate consideration of land and maritime power, not just during operations against enemy forces, but when used as part of a team that protects and aids friendly forces as well.

Much of what airpower can accomplish from within these three domains is done to critically affect events in the land and maritime domains—this is the heart of joint-domain integration, a fundamental aspect of airpower's contribution to US national interests. Airmen integrate capabilities across air, space, and cyberspace domains to achieve effects across all domains in support of Joint Force Commander (JFC) objectives. For example, a remotely piloted aircraft operating from a ground station in the continental US relies on space and cyberspace capabilities to support operations overseas. While all Services rely more and more on such integration, cross-domain integration is fundamental to how Airmen employ airpower to complement the joint force.

Airmen exploit the third dimension, which consists of the entire expanse above the earth's surface. Its lower limit is the earth's surface (land or water), and the upper limit reaches toward infinity. This third dimension consists of the air and space domains. From an operational perspective, the air domain can be described as that region above the earth's surface in which aerodynamics generally govern the planning and conduct of military operations, while the space domain can be described as that region above the earth's surface in which astrodynamics generally govern the planning and conduct of military operations. Airmen also exploit operational capabilities in cyberspace. Cyberspace is a global domain within the information environment consisting of the interdependent network of information technology infrastructures, including the Internet, telecommunications networks, computer systems, and embedded processors and controllers. In contrast to our surface-oriented sister Services, the Air Force uses air, space, and cyberspace capabilities to create effects, including many on land and in the maritime domains, that are ends unto themselves, not just in support of predominantly land or maritime force activities.

The evolution of contemporary airpower stems from the Airman's original vision of combat from a distance, bypassing the force-on-force clash of surface combat. Originally manifest in long-range aircraft delivering kinetic weapons, airpower has evolved over time to include many long-range supporting capabilities, notably the conduct of networked information-related operations. This evolution has accelerated as Airmen conduct a greater percentage of operations not just over-the-horizon but globally, expanding operations first through space and now also in cyberspace. Just as airpower grew from its initial use as an adjunct to surface operations, space and cyberspace have likewise grown from their original manifestations as supporting capabilities into warfighting arenas in their own right.

The Foundations of Airpower

Airpower provides the Nation and the joint force with unique and valuable capabilities. **Airmen should understand the intellectual foundations behind airpower and articulate its proper application at all levels of conflict; translate the benefits of airpower into meaningful objectives and desired effects; and influence the overall operational planning effort from inception to whatever post-conflict operations are required.**

Airpower stems from the use of lethal and nonlethal means by air forces to achieve strategic, operational, and tactical objectives. The Air Force can rapidly provide national leadership and joint commanders a wide range of military options for meeting national objectives and protecting national interests.

Elevation above the earth's surface provides relative advantages and has helped create a mindset that sees conflict more broadly than other forces. Broader perspective, greater potential speed and range, and three-dimensional movement fundamentally change the dynamics of conflict in ways not well understood by those bound to the surface. The result is inherent flexibility and versatility based on greater mobility and responsiveness. Airpower's speed, range, flexibility, and versatility are its outstanding attributes in both space and time. This combination of attributes provides the foundation for the employment concepts of airpower.

With its speed, range, and three-dimensional perspective, **airpower operates in ways that are fundamentally different from other forms of military power.** Airpower has the ability to conduct operations and impose effects throughout an entire theater and across the Range of Military Operations (ROMO), unlike surface forces that typically divide up the battlefield into individual operating areas. Airmen generally view the application of force more from a functional than geographic standpoint, and classify targets by generated effects rather than physical location.

By making effective use of the third dimension, the electromagnetic spectrum, and time, airpower can seize the initiative, set the terms of battle, establish a dominant tempo of operations, better anticipate the enemy through superior observation, and take advantage of tactical, operational, and strategic opportunities. Thus, airpower can simultaneously strike directly at the adversary's centers of gravity, vital centers, critical vulnerabilities, and strategy. Airpower's ability to strike the enemy rapidly and unexpectedly across all of these critical points adds a significant impact to an enemy's will in addition to the physical blow. This capability allows airpower to achieve effects well beyond the tactical effects of individual actions, at a tempo that disrupts the adversary's decision cycle.

Airpower can be used to rapidly express the national will wherever and whenever necessary. Within 36 hours of the deployment order, Air Force F-15s were flying combat air patrols over Saudi Arabia in response to the Iraqi invasion of Kuwait in 1990. More recently, Air Force forces demonstrated that same rapid-response capability by airlifting desperately needed supplies into tsunami-stricken areas of South and Southeast Asia and earthquake-stricken Haiti. The world at large perceives American airpower to be a politically acceptable expression of national power which offers reasonable alternatives to long, bloody ground battles while making an impact on the international situation. While a "boots-on-the-ground" presence may often be required, airpower makes that presence more effective, in less time, and often with fewer casualties. Increasingly, US national power and international influence are gauged in terms of what we can or cannot accomplish with this capability.

The Air Force provides national leadership and joint commanders with options, the threat of which may accomplish political objectives without the application of lethal force. The means is embedded in the ability to respond rapidly to crises anywhere in the world and across the ROMO. An obvious example is the deterrent role played by the Air Force's nuclear-armed bombers and intercontinental ballistic missiles against the Soviet Union during the Cold War. More recently, B-52 and B-2 bombers have rotated into Guam to provide a ready and visible presence.

The Air Force provides the unique ability to hold at risk a wide range of an adversary's options and possible courses of action; this is increasingly the key to successful joint campaigns. Airpower is increasingly the first military instrument brought to bear against an enemy in order to favorably influence the overall campaign. Frequently, and especially during the opening days of a crisis, airpower may be the only military instrument available to use against an enemy; this may be especially true if friendly ground forces are not immediately present in a given region.

Air Force forces can respond rapidly to apply effects. The same spacecraft which Airmen employ to observe hostile territory prior to the outbreak of hostilities provide key intelligence to battle planners. The same aircraft which provide visible deterrence to a potential aggressor can be employed immediately to defend or attack should deterrence fail. The shift from deterrent force to combat power is near-instantaneous. From ready deterrent to bombs-on-target is only a question of command and control and flight time.

Airpower is more than dropping bombs, strafing targets, firing missiles, providing precision navigation and timing, or protecting networks. It is also a way of influencing world situations in ways which support national objectives. To most observers in the post-Cold War world, the use of military power is politically less acceptable than in previous times. This is true even if we act in a purely humanitarian endeavor or influence a given international political situation with a modest show of force. In international disasters, natural or man-made, from the Berlin Airlift to earthquake relief operations in Pakistan, the Air Force is the only military force in the world which has the airlift and air refueling capability to provide immediate relief supplies and personnel in response to global emergencies. Air Force aircraft delivering relief supplies serve not only to alleviate the immediate situation, but also to provide a visible symbol of the care, concern, and capability of the US. Through careful building of partnerships, Air Force forces can favorably shape the strategic environment by assessing, advising, training, and assisting host nation air forces in their efforts to counter internal or external threats. The perception of credible US forces underpins many deterrence and assurance strategies. Such activities lead to greater regional stability and security.

Within the broad sweep of history, the benefits of this instrument of military power are relatively new. Up until the latter part of the 20th century, naval forces provided the primary symbol of American military power and resolve; powerful warships making port calls throughout the world were visible symbols of the strength and capability of the US. Today, airpower plays a very similar role—and not just in those nations with major seaports. In numerous humanitarian operations, Airmen have provided relief, demonstrated resolve, and helped to shape the attitudes of world leaders and their people.

This influence is more than just airplanes. US space-based assets are a non-intrusive method of providing up-to-the-minute warning and information on the maneuver of hostile military forces or other potentially dangerous actions. The US often shares this information with friendly nations in response to potential adversaries to defuse points of conflict before they result in hostilities. US air, space, and cyberspace capabilities provide the means to alert allies of a potential aggressor's hostile intentions or impending attack when in-country physical presence is unwarranted. They can influence potential adversaries by stripping them of the ability to hide hostile military activity without violating national sovereignty.

Airpower's speed, range, flexibility, precision, and lethality provide a spectrum of employment options with effects that range from tactical to strategic. This range of effects is an important contribution. A surface-centric strategy often seeks its outcome through the destruction of hostile land forces and the occupation of territory. However, destruction of hostile land forces may be only a tactical or operational objective and may not achieve the desired strategic outcome. Further, territorial occupation, with its attendant large cultural footprint, may not be feasible or politically acceptable. Sea power, with its ability to project force and disrupt the economic lifeline of a maritime-capable adversary, also provides the potential for strategic results. However, slow surface speeds can constrain its capability to respond rapidly from one theater to another. In addition, it may be extremely vulnerable in littoral regions. Often, in such circumstances, the political risks outweigh the actual military risks.

Airpower, on the other hand, has been successfully used to influence strategic political outcomes in many world crises since the Berlin Airlift of 1948. Throughout the Cold War, and continuing under various international arms control agreements, Air Force assets have been used to observe and verify compliance, leveraging our ability to negotiate and influence diplomatically. If force becomes necessary, Air Force assets can secure strategic outcomes at any time by overflying surface forces and thus bypassing geographical boundaries, or striking with precision at the critical vulnerabilities within an adversary's political, military, and industrial centers of gravity. Even in situations when joint strategy requires large-scale destruction of enemy surface forces, Air Force forces can deliver the bulk of that destruction. It can do these things sooner than can other military forces, and it has been demonstrated that the earlier the application of effects, usually the less total force required. In humanitarian cases, the earlier the relief, the better the effect.

Operating in a seamless medium, there are no natural boundaries to constrain air, space, and cyberspace operations. Through centralized control of Air Force assets and decentralized execution, commanders reap the benefits of airpower throughout the ROMO, wherever most needed at any given time.

Airpower has a degree of versatility not found in any other force. Many aircraft can be employed in a variety of roles and shift rapidly from the defense to the offense. Aircraft may conduct a close air support mission on one sortie, then be rearmed and subsequently used to suppress enemy surface-to-surface missile attacks or to interdict enemy supply routes on the next. In time-sensitive scenarios, aircraft en route to one target, or air mobility aircraft in support of one mission, can be reassigned new targets or re-missioned as new opportunities emerge. Multirole manned and unmanned platforms may perform intelligence, surveillance, and reconnaissance (ISR), command and control (C2), and attack functions all during the same mission, providing more potential versatility per sortie. Finally, aircraft can be repositioned within a theater to provide more responsiveness, while space and cyberspace capabilities can be reprioritized.

Joint campaigns rely upon this versatility. However, many airpower capabilities are limited in number; dividing or parceling out airpower into "penny-packets" violate the tenet of synergy and principle of mass. To preserve unity of effort, JFCs normally vest a single air commander with control of all airpower capabilities.

Historically, armies, navies, and air forces massed large numbers of troops, ships, or aircraft to create significant impact on the enemy. Today, the technological impact of precision guided munitions enables a relatively small number of aircraft to directly achieve national as well as military strategy objectives. When combined with stealth technologies, airpower today can provide shock and surprise without unnecessarily exposing friendly forces. To destroy a single target, we no longer need the thousand-plane bomber raids of World War II or the hundreds of sorties of Vietnam. Today's air forces can provide accurate and assured destruction of vital targets with far fewer aircraft, sometimes multiple targets with a single aircraft. Moreover, that capability can be delivered from within the theater or around the globe if necessary. Whether in the skies of Iraq and Afghanistan, delivering United Nations peacekeeping troops to Africa, or monitoring nuclear weapons proliferation and development, Air Force forces have a far-reaching presence and the ability to produce direct and immediate effects.

With all those characteristics considered, one should remember that **air, space, and cyberspace superiority are the essential first ingredients in any successful modern military operation.** Military leaders recognize that successful military operations can be conducted only when they have gained the required level of control of the domains above the surface domains. Freedom to conduct land and naval operations is substantially enhanced when friendly forces are assured that the enemy cannot disrupt operations from above.

Control of the air, space, and cyberspace domains is not a goal for its own sake, but rather a prerequisite for all other military operations. Air mastery has allowed American land, naval, and air forces to operate where they want, at their own tempo, while creating the environment for success.

"Airmindedness"

The perspective of Airmen is necessarily different; it reflects a unique appreciation of airpower's potential, as well as the threats and survival imperatives unique to Airmen. The study of airpower leads to a particular expertise and a distinctive point of view that General Henry H. "Hap" Arnold termed "airmindedness."

Airmen normally think of airpower and the application of force from a functional rather than geographical

perspective. Airmen do not divide up the battlefield into operating areas as some surface forces do; airmindedness entails thinking beyond two dimensions, into the dimensions of the vertical and the dimension of time. Airmen think spatially, from the surface to geosynchronous orbit. Airmen typically classify targets by the effect their destruction would have on the adversary instead of where the targets are physically located. This approach normally leads to more inclusive and comprehensive perspectives that favor strategic solutions over tactical ones. Finally, Airmen also think of power projection from inside the US to anywhere on the globe in hours (for air operations) and even nanoseconds (for space and cyberspace operations).

Airmindedness impacts Airmen's thoughts throughout all phases of operations. It is neither platform- nor situation-specific. Airmindedness enables Airmen to think and act at the tactical, operational, and strategic levels of war, simultaneously if called for. Thus, the flexibility and utility of airpower is best fully exploited by an air-minded Airman.

The Airman's Perspective

The practical application of "airmindedness" results in the Airman's unique perspective, which can be summarized as follows.

- **Control of the vertical dimension is generally a necessary precondition for control of the surface.** The first mission of an air force is to defeat or neutralize the enemy air forces so friendly operations on land, sea, in the air, and in space can proceed unhindered, while at the same time one's own military forces and critical vulnerabilities remain safe from air attack.

- **Airpower is an inherently strategic force.** War and peace are decided, organized, planned, supplied, and commanded at the strategic level of war. Air Force forces can hold an enemy's strategic centers of gravity and critical vulnerabilities directly at risk immediately and continuously. Airpower also has great strategic capability for non-lethal strategic influence, as in humanitarian relief and building partnership activities.

- **Airpower can exploit the principles of mass and maneuver simultaneously to a far greater extent than surface forces.** There are no natural lateral boundaries to prevent air, space, and cyberspace capabilities from quickly concentrating their power (physically or in terms of delivered effects) at any point, even when starting from widely dispersed locations. Airpower dominates the fourth dimension—time—and compresses the tempo of events to produce physical and psychological shock.

- **Airpower can apply force against many facets of enemy power.** Air Force-provided capabilities can be brought to bear against any lawful target within an enemy's diplomatic, informational, military, economic, and social structures simultaneously or separately. They can be employed in support of national, combined/joint, or other component objectives. They can be integrated with surface power or employed independently.

- **Air Force forces are less culturally intrusive in many scenarios.** Surface forces are composed of many people and vehicles which, when arrayed for operations, cover a significant area. Thus, their presence may be very visible to local populations and may create resentment during certain types of stability operations and in counterinsurgency operations. Air Force forces, operating from bases over the horizon or from just a few bases in-country, have a smaller footprint for the effects they provide. Space and cyberspace forces have a negligible in-theater footprint relative to the capabilities they provide.

- **Airpower's inherent speed, range, and flexibility combine to make it one of the most versatile components of military power.** Its versatility allows it to be rapidly employed against strategic, operational, and tactical objectives simultaneously. The versatility of airpower derives not only from the inherent characteristics of air forces themselves, but also from the manner in which they are organized and controlled.

- **Airpower results from the effective integration of capabilities, people, weapons, bases, logistics, and all supporting infrastructure.** No one aspect of air, space, and cyberspace capabilities should be treated in isolation since each element is essential and interdependent. Ultimately, the Air Force depends on the performance of the people who operate, command, and sustain air, space, and cyberspace forces.

- **The choice of appropriate capabilities is a key aspect in the realization of airpower**. Weapons should be selected based on their ability to create desired effects on an adversary's capability and will. Achieving the full potential of airpower requires timely, actionable intelligence and sufficient command and control capabilities to permit commanders to exploit precision, speed, range, flexibility, and versatility.

- **Supporting bases with their people, systems, and facilities are essential to launch, recovery, and sustainment of Air Force forces.** One of the most important aspects of the Air Force has proved to be its ability to move anywhere in the world quickly and then rapidly begin operations. However, the need for mobility should be balanced against the need to operate at the deployment site. The availability and operability of suitable bases can be the dominant factor in employment planning and execution.

- **Airpower's unique characteristics necessitate that it be centrally controlled by Airmen.** Airpower can quickly intervene anywhere, regardless of whether it is used for strategic or tactical purposes. Thus, Airmen tend to take a broader view of war, because the capabilities they command have effects at broader levels of war. Airmen apply airpower through the tenet of centralized control and decentralized execution.

CORE FUNCTIONS

A modern, autonomous, and thoroughly trained Air Force in being at all times will not alone be sufficient, but without it there can be no national security.

— General H. H. "Hap" Arnold

Recently the Air Force refined its understanding of the core duties and responsibilities it performs as a Service, streamlining what previously were six distinctive capabilities and seventeen operational functions into twelve core functions to be used across the doctrine, organization, training, materiel, leadership and education, personnel, and facilities spectrum. These core functions express the ways in which the Air Force is particularly and appropriately suited to contribute to national security, but they do not necessarily express every aspect of what the Air Force contributes to the nation.

- Nuclear Deterrence Operations
- Air Superiority
- Space Superiority
- Cyberspace Superiority
- Command and Control
- Global Integrated ISR
- Global Precision Attack
- Special Operations
- Rapid Global Mobility
- Personnel Recovery
- Agile Combat Support
- Building Partnerships

Nuclear Deterrence Operations

The purpose of Nuclear Deterrence Operations is to operate, maintain, and secure nuclear forces to achieve an assured capability to deter an adversary from taking action against vital US interests. In the event deterrence fails, the US should be able to appropriately respond with nuclear options. The sub-elements of this function are:

- Assure/Dissuade/Deter
- Nuclear Strike
- Nuclear Surety

Air Superiority

Air Superiority is that degree of dominance in the air battle of one force over another which permits the conduct of operations by the former and its related land, sea, air, and special operations forces at a given time and place without prohibitive interference by the opposing force. The sub-elements of this function are:

- Offensive Counterair
- Defensive Counterair
- Airspace Control

Space Superiority

Space superiority is the degree of dominance in space of one force over another that permits the conduct of operations by the former and its related land, sea, air, space, and special operations forces at a given time and place without prohibitive interference by the opposing force. Space superiority may be localized in time and space, or it may be broad and enduring. Space superiority provides freedom of action in space for friendly forces and, when directed, denies the same freedom to the adversary. The sub-elements of this function are:

- Space Force Enhancement
- Space Force Application
- Space Control
- Space Support

Cyberspace Superiority

Cyberspace Superiority is the operational advantage in, through, and from cyberspace to conduct operations at a given time and in a given domain without prohibitive interference. The sub-elements of this function are:

- Cyberspace Force Application
- Cyberspace Defense
- Cyberspace Support

Command and Control

Command and control is the exercise of authority and direction by a properly designated commander over assigned and attached forces in the accomplishment of the mission. Command and control functions are performed through an arrangement of personnel, equipment, communications, facilities, and procedures employed by a commander in planning, directing, coordinating, and controlling forces and operations in the accomplishment of the mission. This core function includes all of the C2-related capabilities and activities associated with air, space, cyberspace, nuclear, and agile combat support operations to achieve strategic, operational, and tactical objectives.

Global Integrated Intelligence, Surveillance, and Reconnaissance

Global Integrated ISR is the synchronization and integration of the planning and operation of sensors, assets, and processing, exploitation, dissemination systems across the globe to conduct current and future operations. The sub-elements of this function are:

- Planning and Directing
- Collection
- Processing and Exploitation
- Analysis and Production
- Dissemination and Integration

Global Precision Attack

Global Precision Attack is the ability to hold at risk or strike rapidly and persistently, with a wide range of munitions, any target and to create swift, decisive, and precise effects across multiple domains. The sub-elements of this function are:

- Strategic Attack
- Air Interdiction
- Close Air Support

Special Operations

Special Operations are operations conducted in hostile, denied, or politically sensitive environments to achieve military, diplomatic, informational, and/or economic objectives employing military capabilities for which there is no broad conventional force requirement. These operations may require covert, clandestine, or low-visibility capabilities. Special operations are applicable across the ROMO. They can be conducted independently or in conjunction with operations of conventional forces or other government agencies and may include operations through, with, or by indigenous or surrogate forces. Special operations differ from conventional operations in degree of physical and political risk, operational techniques, mode of employment, independence from friendly support, and dependence on detailed operational intelligence and indigenous assets. The sub-elements of this function are:

- Agile Combat Support
- Aviation Foreign Internal Defense
- Battlefield Air Operations
- Command and Control
- Information Operations
- Intelligence, Surveillance, and Reconnaissance
- Military Information Support Operations
- Precision Strike
- Specialized Air Mobility
- Specialized Refueling

Rapid Global Mobility

Rapid Global Mobility is the timely deployment, employment, sustainment, augmentation, and redeployment of military forces and capabilities across the ROMO. It provides joint military forces the capability to move from place to place while retaining the ability to fulfill their primary mission. Rapid Global Mobility is essential to virtually every military operation, allowing forces to reach foreign or domestic destinations quickly, thus seizing the initiative through speed and surprise. The sub-elements of this function are:

- Airlift
- Air Refueling
- Aeromedical Evacuation

Personnel Recovery

Personnel Recovery (PR) is defined as the sum of military, diplomatic, and civil efforts to prepare for and execute the recovery and reintegration of isolated personnel. It is the ability of the US government and its international partners to affect the recovery of isolated personnel across the ROMO and return those personnel to duty. PR also enhances the development of an effective, global capacity to protect and recover isolated personnel wherever they are placed at risk; deny an adversary's ability to exploit a nation through propaganda; and develop joint, interagency, and international capabilities that contribute to crisis response and regional stability. The sub-elements of this function are:

- Combat Search and Rescue
- Civil Search and Rescue
- Disaster Response
- Humanitarian Assistance Operations
- Medical Evacuation/Casualty Evacuation

Agile Combat Support

Agile Combat Support is the ability to field, protect, and sustain Air Force forces across the ROMO to achieve joint effects. The sub-elements of this function are:

- Ready the Total Force
- Prepare the Battlespace
- Position the Total Force
- Protect the Total Force
- Employ Combat Support Forces
- Sustain the Total Force
- Recover the Total Force

Building Partnerships

Building Partnerships is described as Airmen interacting with international airmen and other relevant actors to develop, guide, and sustain relationships for mutual benefit and security. Building Partnerships is about interacting with others and is therefore an inherently inter-personal and cross-cultural undertaking. Through both words and deeds, the majority of interaction is devoted to building trust-based relationships for mutual benefit. It includes both foreign partners as well as domestic partners and emphasizes collaboration with foreign governments, militaries and populations as well as US government departments, agencies, industry, and non-governmental organizations (NGOs). To better facilitate partnering efforts, Airmen should be competent in the relevant language, region, and culture. The sub-elements of this function are:

- Communicate
- Shape

CONCLUSION

If there is one attitude more dangerous than to assume that a future war will be just like the last one, it is to imagine that it will be so utterly different that we can afford to ignore all the lessons of the last one.

— Air Marshall Sir John C. Slessor

More and more often, our national leadership is calling upon airpower as the military instrument of first choice, and they are asking it to accomplish tasks previously held unworkable—to coerce and to compel. Airpower offers joint force commanders options, including the ability to go to the heart of an enemy and attain a variety of effects directly at the strategic level. To support our national leadership, Airmen, as military professionals, must think about how to accomplish a spectrum of missions. We must understand the potential of airpower, and be able to plan and employ it to its maximum effect, and to articulate it within the context of joint operations. This is especially true in contemporary irregular warfare operations, in which airpower plays an important role, but largely complementing surface operations.

Air Force doctrine development is never totally complete—it is a continuous work in progress. We must remain aware of the lessons of the past—alert and receptive to future technologies and paradigms that may alter the art of air, space, and cyberspace warfare. We should not assume that things have not or will not change; above all, doctrine should be continually interpreted in light of the present situation. A too-literal reading of doctrine may fail to accommodate new operational realities.

Doctrine application requires informed judgment. Certain principles—like unity of command, objective, and offensive—have stood the test of time. Other ideas—like unescorted daytime bombing, decentralized command, and the preeminence of nuclear weapons—have not. If we ignore the potential of integrated air, space, and cyberspace operations and the global and strategic potential of airpower, we may commit the same sins as our forebears by preparing for the "wrong war." If we ignore the reality that adaptive, thinking adversaries will seek asymmetric strategies, anti-access capabilities, and favorable arenas within which to influence and engage us, we risk failure. Tomorrow, a new set of conditions and requirements will likely emerge. In fact, some new conditions and environments are already emerging, and national security requirements are changing. The best hedge is an institutional commitment to learn from experience and to exploit relevant ideas and new technologies so we may be ready for the future, while retaining those fundamental principles that remain constant over time.

14.5 Should the US Maintain the Nuclear Triad?

By Dr. Adam B. Lowther

OBJECTIVES:

20. Name the components of the nuclear triad.
21. Describe the background and intent of President Eisenhower's "New Look" policy.
22. Define the concept of "assured destruction."
23. In your own terms, relate the author's stated reasons for keeping the nuclear triad in place.

In the first week of Pres. Barack Obama's new administration, the White House released his agenda, stating the policies the president will pursue regarding the nuclear arsenal. The agenda includes three foci: securing loose nuclear material from terrorists, strengthening the Nuclear Non- Proliferation Treaty, and moving toward a nuclear-free world.[1] Pushing the president in the direction of a "world without nuclear weapons" are such paragons of past political power as former senator Sam Nunn and former secretaries of state George Shultz and Henry Kissinger.[2] Adding a host of Washington's think-tank analysts to this list produces a crescendo of voices calling for "global zero." They challenge not only the current size of the arsenal but also the very need for a nuclear triad. Much of the recent scholarship shows a clear preference for moving to a monad composed solely of submarines armed with submarine-launched ballistic missiles (SLBM) until the United States ultimately disarms.[3]

Some past and present members of the military leadership hold a view that supports the nuclear arsenal. Senior leaders have given a number of public speeches and interviews outlining what it will take to maintain and modernize the most advanced and secure nuclear arsenal in the world.[4] A key aspect of the general position held by supporters of the arsenal includes retaining the triad and replacing aging platforms.

In the ongoing debate over the appropriate size and purpose of the nuclear arsenal, abolitionists—clearly in the ascendency— make six basic arguments that would ultimately lead to creation of a nuclear monad before reaching total disarmament:[5]

1. Post–Cold War presidents have failed to alter nuclear policy for the current security environment.

2. Terrorism, not Russia, is the primary threat facing the United States. Nuclear weapons do not deter terrorists.

3. America's advanced conventional capabilities can accomplish the same objectives as nuclear weapons.

4. As a signer of the Nuclear Non-Proliferation Treaty, the United States must move toward nuclear abolition.

5. Only nuclear disarmament can overcome the threats of accidental detonation, miscalculation leading to nuclear war, and proliferation of nuclear weapons and material.

6. The safest and most secure leg of the nuclear triad is the sea-based one. Thus, it should become the sole delivery platform for the nuclear arsenal.[6]

Admittedly, each of these arguments has some element of truth; they do not, however, represent a complete understanding of the strategic role played by nuclear weapons in ensuring the sovereignty of the United States or the specific contribution of each leg of the triad. Although each of the abolitionists' arguments deserves a detailed refutation, a focus on the relevance of the triad must suffice.

DEVELOPMENT OF THE TRIAD

In 1947, the year the United States Air Force became an independent service, the American military was attempting to develop sound tactical, operational, and strategic doctrine for the use of nuclear weapons. Just two years earlier, a new and devastating weapon had changed the face of warfare, but the full implications of the atom bomb were yet to be realized. In a flurry of activity, the academic, military, and policy communities undertook much writing and studying as the nation sought to understand nuclear weapons while also confronting the Soviet Union. As technology developed over the following decades, the nation moved from depending on a fleet of long-range bombers as the sole method of delivering nuclear weapons (1945–59) to a nuclear triad composed of bombers, intercontinental ballistic missiles (ICBM), and SLBMs.[7]

During the 1950s, Pres. Dwight Eisenhower believed that an American effort to maintain conventional parity with the Soviet Union would destroy the US economy and bankrupt the federal treasury.[8] Thus, his administration turned to the nuclear arsenal as a substitute for conventional parity. In the president's view, the United States could effectively deter Soviet aggression by placing greater emphasis on nuclear weapons in American national security policy. Commonly called the "New Look," the president's emphasis on the growth of advanced nuclear weapons and delivery platforms led to development of a large fleet of nuclear bombers and, by the end of the Eisenhower administration, the nuclear triad.[9] Composed of three legs, the triad provides the United States with three distinct delivery platforms for nuclear weapons.

The first and oldest leg includes the nation's long-range bombers and their payload of gravity bombs and air launched cruise missiles. At its apex in the early to mid-1960s, Strategic Air Command included more than 1,300 nuclear-capable bombers, including 700 of the then-new B-52s.[10] By 1990 the nation's long-range bomber fleet had declined to 347 total aircraft.[11] Today, nuclear-capable bombers account for about half of the Air Force's bomber fleet of 162 aircraft.[12]

A second leg became part of the nation's nuclear arsenal in 1959 with deployment of the first six Atlas D ICBMs. Just three years later, the first Minuteman I deployed. Not until 1970 did America's ICBM force reach its peak with a mix of 1,054 Titan II and Minuteman I, II, and III missiles—most of which carried three to 12 warheads. These numbers remained constant until 1982.[13] Since then, the number of operationally deployed ICBMs has steadily declined to its current size of 450.[14]

The addition of the Polaris SLBM in 1960 completed the triad. Like the other two legs, SLBMs waxed at the height of the Cold War and waned as it ended. By 1967 the United States had deployed 656 SLBMs aboard 41 ballistic missile submarines (SSBN). When the Soviet Union collapsed in December 1991, the sea leg of the triad remained largely intact with 33 SSBNs carrying 608 SLBMs.[15] Today, however, only 14 *Ohio*-class submarines remain, each carrying 24 Trident II nuclear missiles.

Throughout the Cold War, the United States maintained a substantial inferiority in conventional military forces but enjoyed the protection of a sizable nuclear umbrella. As the Cold War progressed and American thinking about nuclear conflict developed, "assured destruction" took precedence as the approach of choice. Developed by Thomas Schelling and others while he worked for the RAND Corporation in the 1960s, the concept of assured destruction purposefully left the United States vulnerable to a first strike, yet the nation maintained a credible second-strike capability.[16] Although nuclear policy evolved throughout the Cold War, its essential nature remained much the same. Because of the exorbitant fiscal cost of building a large underground industrial infrastructure, for example, the nation chose to accept the risk of an unprotected public—but only as long as it was defended by bombers standing at alert, ICBMs protected in their reinforced silos, and submarines quietly prowling the world's oceans. In the end, deterrence seems to have worked.

A second aspect of American nuclear policy—often overlooked in the current debate— dates back to the earliest days of the North Atlantic Treaty Organization (NATO) when the United States and its European allies made a conscious decision to forgo creation of a NATO military equal in strength to that of the Warsaw Pact. Instead, the European members of NATO chose to rely on America's strategic nuclear weapons— based in the United States and at sea— as well as tactical nuclear weapons, based in Europe, as a guarantor that Eastern Bloc troops would not roll through the Fulda Gap on their way to Paris.[17] Extended deterrence, as it came to be known, enabled Western Europe to focus on economic development instead of heavy investment in national security. Although this type of deterrence often proved unpopular with European publics, governments throughout Western Europe depended upon the security provided by basing nuclear weapons throughout the West.

ENTERING THE POST–COLD WAR ERA

In the immediate aftermath of the Cold War, assured destruction and related nuclear strategies that had served the nation well for more than two generations were almost forgotten as the euphoria that engrossed America took hold.[18] With it, the triad fell into decline. As the former Soviet Union sought to stabilize its deteriorating economy by lowering its military expenditures, the United States joined Russia in making dramatic reductions to the overall size of the nuclear arsenal. The "peace dividend" promised to the American people by presidents George H. W. Bush and Bill Clinton led to a refocusing of US foreign policy. With the Russian Bear focused on internal struggles, the United States was free to take on the role of global hegemon and concentrate its efforts on serving as the world's policeman. The 1990s saw the US military intervene in a number of failing or failed states such as Somalia, Haiti, Bosnia, and Serbia, while also emphasizing democratization of the former Soviet Union and globalization of the international economy.[19]

As Francis Fukuyama suggested in his article "The End of History?" "What we may be witnessing is not just the end of the Cold War, or the passing of a particular period of postwar history, but the end of history as such: that is, the end point of mankind's ideological evolution and the universalization of Western liberal democracy as the final form of human government."[20] Democracy had apparently won; socialism had apparently lost. Continuing to focus on the nuclear triad and nuclear conflict seemed passé.

Between 1991 and 2009, the nuclear arsenal shrank by more than 75 percent. Few members of Congress or the military objected since it appeared that the single greatest purpose for nuclear weapons was gone. Even in the wake of the terrorist attacks of 11 September 2001, Pres. George W. Bush signed the Strategic Offensive Reduction Treaty, which obligates the United States and Russia to reduce their operationally deployed strategic weapons to between 1,700–2,200 each by 2012. President Obama is promising to follow suit and continue reductions in the nuclear arsenal as the United States eventually moves to zero.[21]

Although President Obama's speech of 5 April 2009 may give the impression that he has adopted the stance of nuclear abolitionists, one should not forget that Pres. Ronald Reagan once said that he "dream[ed]" of a "world free of nuclear weapons."[22] Just as Reagan shepherded the United States to victory in the Cold War, so, hopefully, will President Obama act responsibly and not put the national security of the United States at risk by reducing the nuclear arsenal to a point that nuclear deterrence loses the credibility that enables its success.

THE CURRENT DEBATE

In an era dominated by non-state actors (terrorists, international criminal gangs, and insurgents), rogue regimes, and rising powers, some members of the Air Force are asking whether the triad is still relevant or whether nuclear abolitionists are correct in suggesting that the United States adopt a monad as the nation moves toward zero. The answers to these questions deserve considerable attention. In short, however, the triad is as relevant today as it was at the height of the Cold War. Nevertheless, before offering a justification for maintaining the triad, one should explain the position of nuclear abolitionists.

The Abolitionists' Position

According to the most recent reports and studies published by advocates of nuclear abolition, the United States should initiate complete disarmament by taking the following actions.[23] First, abolitionists desire to remove the 76 remaining B-52H and 19 B-2 bombers from nuclear-capable service.[24] By maintaining an arsenal of 500–1,000 warheads, as abolitionists suggest, the United States no longer needs the bomber leg of the triad. Additionally, the nation's long-range bombers are slow to reach their targets, cannot penetrate advanced anti-air defenses (with the exception of the B-2), and are expensive to procure and maintain.

Second, abolitionists seek to dismantle the nation's 450 ICBMs, which need expensive upgrades or replacement and present the nation's adversaries a target on US soil.

Third, abolitionists are willing to accept, for the near term, a nuclear deterrence strategy that relies solely on a dozen Ohio-class SSBNs (after downsizing from the present 14), each armed with 24 Trident II SLBMs.[25] According to their strategy, the United States will maintain half of its SSBNs at sea at any given time while the other half is in port at one of two designated submarine bases.

Abolitionists are willing to accept a submarine-based monad because they consider submarines the most secure leg of the triad. These vessels also obviate the need for operationally deployed nuclear weapons on US soil. Supposedly, the absence of these weapons would reduce the likelihood of a counterforce strike against the homeland.

Because these arguments seem reasonable and each contains an element of truth, they have wide appeal. But if the United States were to adopt a monad, the nation's ability to deter current and future adversaries would decline precipitously for four key reasons.

The Counterview

First, deterrence, the capstone of American foreign policy since the end of World War II, relies on effectively making an adversary believe that the risks involved in changing the status quo outweigh any potential rewards. To achieve effective deterrence, the United States must have the capability and, most importantly, credibility to create the desired psychological effect. Moving to a nuclear deterrence strategy that effectively depends on a half dozen deployed submarines undermines both capability and credibility. Contrary to the admonitions of abolitionists, adopting a monad sends a clear signal to America's adversaries that the nation does not value nuclear weapons to the degree it once did and will be more reluctant to use a diminished arsenal in the future. This emboldens adversaries and decreases the confidence that US allies have in the nation's extended deterrence.

Successful deterrence depends *completely* upon simply and effectively communicating desire and intent to allies and adversaries. Dramatically reducing the size of the arsenal and killing two legs of the triad, while claiming that the United States remains serious about nuclear deterrence, would send a mixed signal. The historical record does not offer analogous examples of arms reductions leading to the maintenance of credibility. On the contrary, the Washington Naval Treaty (1922), which limited the tonnage of major world navies, may have played a key role in leading the Japanese to attack Pearl Harbor.[26] Admittedly, such counterfactual claims are difficult to prove.

Second, since signaling intent is a vital aspect of successful deterrence, eliminating the bomber leg of the triad would be a mistake. Designed to remain hidden from the view of an adversary, ICBMs and SSBNs offer no effective way of conveying American resolve or an escalation/de-escalation in posture, should an adversary move toward conflict. The bomber fleet, however, effectively demonstrates resolve. For example, if an adversary were to openly challenge the status quo, the president could order the nation's B-52s and B-2s on alert, put them in the air, and/or deploy them to forward bases. All of these actions are visible signals of American intent, designed to lead to a de-escalation of tensions. Without question, bombers are the most effective tool for overtly demonstrating resolve.

A related point arises. Nuclear-capable bombers are one of the best tools for assuring allies that the United States remains committed to providing a credible extended deterrent. Neither ICBMs nor submarines can provide a visible show of resolve in the face of danger. Deploying nuclear bombers to an ally's air base not only assures America's friends but also deters the nation's foes.

Third, ICBMs offer two distinct benefits that a submarine force cannot replicate. On the one hand, they raise the cost of entry into the nuclear club as a peer of the United States. ICBMs require expensive and advanced missile technology, which may prove too costly for many potential proliferators. On the other hand, they increase risks for an adversary by driving him to a strategy (counterforce) requiring the elimination of American ICBMs in an effort to prevent a US counterstrike. Forcing an adversary to strike the United States in order to eliminate its nuclear arsenal serves as a strong deterrent when the enemy considers a nuclear attack. Moreover, these missiles are the only leg of the triad that can hit any spot on the earth within half an hour.

Fourth, should the United States adopt the plan advocated by abolitionists, the nation's adversaries would know full well that half the nuclear arsenal would be in port at any given time, vulnerable to destruction by a single nuclear missile targeting each of the two designated nuclear submarine bases. Contrary to what Americans are led to believe, Russia and China maintain advanced submarine-detection capabilities that may enable either nation to detect, track, and sink the half of the nuclear arsenal (six submarines) at sea.[27] Moving to a submarine-based monad will also encourage adversaries of the United States to focus technological development on advanced sonar and torpedo technology. Doing so will simplify the calculation for an adversary seeking to neutralize the American arsenal.

The United States may soon face a real scenario in which two nuclear missiles and a half dozen torpedoes can destroy the entire operationally deployed strategic nuclear arsenal—something no American should desire. Redundancy, which the triad provides, offers a level of protection that a submarine-based nuclear arsenal would greatly diminish.

Increasing American vulnerability and decreasing American capability do not represent a strategy for successful deterrence. As history demonstrates, deterrence works when the United States effectively convinces its adversaries that an attack on America will fail to carry out the desired objectives and will invoke massive retaliation. Any other approach to deterrence is doomed to failure.

Relying on what abolitionists refer to as "minimum deterrence" is a recipe for placing the American people at greater risk, not less.[28] Even though the United States will likely suffer a terrorist attack, it is certainly not the most dangerous threat the nation faces. With the nuclear club expanding and likely to gain new members hostile to the United States, weakening the nuclear triad is unwise. Doing so not only will undermine American credibility but also will cause allies to doubt America's commitment to extended deterrence. This could lead allies to pursue their own nuclear arsenals as a hedge against American weakness and perceived threats yet to materialize.

Even though we Americans are generous, well-intentioned people, others do not necessarily wish us well. We would be wise to remember that fact. As the great Roman strategist Vegetius once wrote, "Si vis pacem para bellum" (If you desire peace, prepare for war).

NOTES

1. Barack Obama, "Remarks by President Barack Obama," Office of the Press Secretary, The White House, 5 April 2009, http://www.whitehouse.gov/the_press_office/Remarks-By-President-Barack-Obama-In-Prague-As-Delivered.

2. Ibid.; and George P. Shultz et al., "Toward a Nuclear-Free World," Wall Street Journal, 15 January 2008, http://online.wsj.com/public/article_print/SB120036422673589947.html.

3. Ivo Daalder and Jan Lodal, "The Logic of Zero," Foreign Affairs 87, no. 6 (November 2008): 80–95.

4. Gen Kevin P. Chilton (remarks to the Strategic Weapons in the 21st Century Conference, Ronald Reagan International Trade Center, Washington, DC, 31 January 2008); and Office of the Under Secretary of Defense for Acquisition, Technology, and Logistics, Report of the Defense Science Board Task Force on Nuclear Deterrence Skills (Washington, DC: Office of the Under Secretary of Defense for Acquisition, Technology, and Logistics, September 2008).

5. Within the group broadly defined as nuclear abolitionists are a number of varying opinions. Some—such as Richard Branson and Queen Noor of Jordan, who ascribe to global zero—support a unilateral move by all nuclear powers to abolish nuclear weapons. Others, such as Henry Kissinger and George Shultz, believe that slow and steady reductions are the proper approach. All parties within the abolitionist camp believe in nuclear abolition as a relevant and

obtainable goal. See the Global Zero Web site, http://www.globalzero.org; and Barack Obama, "Remarks by the President at the United Nations Security Council Summit on Nuclear Non-Proliferation and Nuclear Disarmament," Office of the Press Secretary, The White House, 24 September 2009, http://www.whitehouse.gov/the_press_office/Remarks-By-The-President-At-the-UN-Security-Council-Summit-On-Nuclear-Non-Proliferation-And-Nuclear-Disarmament.

6. For a detailed discussion of the arguments made by nuclear abolitionists, see Adam Lowther, Challenging Nuclear Abolition, Research Paper 2009-4 (Maxwell AFB, AL: Air Force Research Institute, August 2009), http://www.afa.org/EdOp/2010/Logic_of_Nuclear_Arsenal.pdf.

7. Douglas P. Lackey, Moral Principles and Nuclear Weapons (New York: Rowman and Littlefield, 1986), 43.

8. Douglas Kinnard, President Eisenhower and Strategy Management: A Study in Defense Politics (New York: Pergamon-Brassey's, 1989), 1-25.

9. Saki Dockrill, Eisenhower's New-Look National Security Policy, 1953-61 (New York: St. Martin's Press, 1996), 48-62.

10. Rebecca Grant, Return of the Bomber: The Future of Long-Range Strike, Air Force Association Special Report (Arlington, VA: Air Force Association, February 2007).

11. Natural Resources Defense Council, "Table of US Strategic Bomber Forces" (Washington, DC: Natural Resources Defense Council, 2002), http://www.nrdc.org/nuclear/nudb/datab7.asp#ninety.

12. "2009 USAF Almanac: The Air Force in Facts and Figures," Air Force Magazine 92, no. 5 (May 2009): 48.

13. Natural Resources Defense Council, "Table of US ICBM Forces" (Washington, DC: Natural Resources Defense Council, 2002), http://www.nrdc.org/nuclear/nudb/datab3.asp.

14. "LGM-30G Minuteman III," fact sheet, US Air Force, Dec. 2009, http://www.af.mil/information/factsheets/factsheet.asp?id=113.

15. Natural Resources Defense Council, "Table of US Ballistic Missile Submarine Forces" (Washington, DC: Natural Resources Defense Council, 2002), http://www.nrdc.org/nuclear/nudb/datab5.asp.

16. See Thomas C. Schelling, Arms and Influence (New Haven, CT: Yale University Press, 1966). In this seminal work on coercion, Schelling lays out the concepts that served as the rationale for Cold War deterrence strategy.

17. David S. Painter, The Cold War: An International History (New York: Routledge, 1999); and Stephen J. Cimbala, The Past and Future of Nuclear Deterrence (Westport, CT: Praeger Publishing, 1998), 11-12, 23-25.

18. Charles Krauthammer, "Don't Cash the Peace Dividend," Time Magazine, 26 March 1990, http://www.time.com/time/magazine/article/0,9171,969672,00.html; and Keith B. Payne, The Great American Gamble: Deterrence Theory and Practice from the Cold War to the Twenty-first Century (Fairfax, VA: National Institute Press, 2008), chap. 3.

19. See Joseph E. Stiglitz, Globalization and Its Discontents (New York: W. W. Norton, 2002), chap. 5.

20. Francis Fukuyama, "The End of History?" National Interest, no. 16 (Summer 1989): 4.

21. Obama, "Remarks by President Barack Obama."

22. Paul Lettow, Ronald Reagan and His Quest to Abolish Nuclear Weapons (New York: Random House, 2005), 6.

23. Hans M. Kristensen, Robert S. Norris, and Ivan Oelrich, From Counterforce to Minimal Deterrence:

A New Nuclear Policy on the Path toward Eliminating Nuclear Weapons, Occasional Paper no. 7 (Washington, DC: Federation of American Scientists and the Natural Resources Defense Council, April

2009), http://www.fas.org/pubs/_docs/OccasionalPaper7.pdf; and Joint Working Group of the American Association for the Advancement of Science; American Physical Society; and Center for Strategic and International Studies, Nuclear Weapons in 21st Century U.S. National Security (Washington, DC: AAAS, APS, and Center for Strategic & International Studies, December 2008), http://www.aps.org/policy/reports/popa-reports/upload/nuclear-weapons.PDF.

24. Sidney D. Drell and James E. Goodby, What Are Nuclear Weapons For? Recommendations for Restructuring U.S. Strategic Nuclear Forces (Washington, DC: Arms Control Association, October 2007), http://www.armscontrol.org/system/files/20071104_Drell_Goodby_07_new.pdf.

25. This is an acknowledgement by the more pragmatic members of the abolitionist camp that unilateral disarmament is not possible. See George P. Shultz et al., "How to Protect Our Nuclear Deterrent," Wall Street Journal, 19 January 2010, http://online.wsj.com/article/SB10001424052748704152804574628344282735008.html.

26. Had the Pacific fleet been as large as advocated by the Department of the Navy in the years prior to 7 December 1941 and not constrained by an arms-limitation treaty, there is reason to believe that the Japanese would not have come to the conclusion that a "knockout blow" was possible.

27. Marcel van Leeuwen, "Russia Starts Ka-28 ASW Deliveries to China," Aviation News, 11 October 2009, http://www.aviation-news.eu/2009/10/11/russia-starts-ka-28-asw-deliveries-to-china/; and "Russian Sonar Technology," Warfare, http://warfare.ru/?linkid=2085&catid=332 (accessed 29 January 2010).

28. Kristensen, Norris, and Oelrich, From Counterforce to Minimal Deterrence, 1-2.

ABOUT THE AUTHOR

Dr. Adam B. Lowther is a military defense analyst with the Air Force Research Institute, Maxwell AFB, Alabama.

From: Dr. Adam B. Lowther, "Should the United States Maintain the Nuclear Triad?," Air & Space Power Journal (Summer 2010): 23-29. Used with permission.

This page has been intentionally left without meaningful content.

15

CHAPTER 15

ORGANIZATIONAL CULTURE & CHANGE

Because organizations are made up of individuals with different talents, personalities, and goals, the organization will have a distinct culture. Some aspects of this culture change when the personnel do; other aspects seem to be fixed and enduring. The anatomy of an organization's culture – how the business functions on a day-to-day base – can strongly influence that organization's potential for success or failure. In addition, the ability of an organization and its leaders to cope with change and encourage innovation also impacts mission effectiveness.

INTRODUCTION

Have you ever entered a new organization and tried to make a change, only to be told, "We tried that and it never works"? Or asked about a process and been told, "That's the way we've always done it"? Inertia and habit are hard for an organization to break. But a struggling squadron with a poor organization culture doesn't have to fold – it can change. You were introduced to the concept of organizational culture and change in chapter 11. Read the first article in this chapter, "Organizational Culture," to identify a common framework that is necessary for changing an organization's culture. The author suggests that a process of engagement, cycle, and review is one of the best ways to recognize the characteristics that define an organization.

Since there's no way to stop change from happening, you and your organization will fare better if you can take change in stride and adapt without complaint. If you have a big change coming up – at school, home, work, or CAP – try the methods listed in the second article, "Manage Change – Not the Chaos Caused by Change," to ease the process. The author identifies several positive steps to make a change program successful, including opening channels of communication, developing a learning environment, and providing training.

Even with open communication, careful planning, and extensive training, your new program or idea may still meet with resistance. The author of "Keeping Change on Track" explains why change can be so difficult, and lists a number of pitfalls that derail change efforts. Knowing about them in advance can help you watch for and avoid common problems during times of change.

In our current environment of global communication, rapid change, and instant access to information, innovation can be crucial to an organization's survival. Finding ways to encourage creativity and innovative work within your team, staff, or squadron can be a challenge for any leader. The fourth article, "Developing an Innovative Culture," reveals several ways to improve culture in organizations through strategies that involve soliciting feedback from employees, encouraging open communication across companies, and encouraging new ideas. Organizations should take advantage of opportunities that arise for learning and development when employees change positions, and leaders can encourage innovation by setting an example of trust and by sharing time and experience with employees. As you read this article, try replacing the word 'employees' with 'volunteers' to see how the author's message applies to CAP.

The final article in this chapter points out personal leadership traits to develop in order to help your organization develop a positive culture and deal with change. In "The Twenty-First Century Leader: Social Artist, Spiritual Visionary, and Cultural Innovator," the author suggests that the increasing complexity and rapidity of change in the modern day calls for a new type of leader, one who can combine art, vision, and innovation.

CHAPTER OUTLINE
This chapter's readings are:

Organizational Culture
Dorian LaGuardia, "Organizational Culture," *T+D* 62, no. 3 (March 2008): 56-61.

Managing Change – Not the Chaos Caused by Change
Beverly Goldberg, "Manage Change – Not the Chaos Caused by Change," *Management Review* 81, no. 11 (November 1992): 39-45.

Keeping Change on Track
Richard Bevan, "Keeping Change on Track," *The Journal For Quality & Participation* 34, no. 1 (April 2011): 4-9.

Developing an Innovative Culture
Erika Agin and Tracy Gibson, "Developing an Innovative Culture," *T+D* 64, no. 7 (July 2010): 52-55.

The Twenty-First Century Leader
Fahri Karakas, "The Twenty-First Century Leader: Social Artist, Spiritual Visionary, and Cultural Innovator," *Global Business & Organizational Excellence*, (March/April 2007): 44-50.

CHAPTER GOALS

1. Appreciate the role culture has in organizational effectiveness.
2. Appreciate the need for leaders to affect change in organizational culture
3. Describe principles of managing cultural change in a positive way.

15.1 Organizational Culture

By Dorian LaGuardia

OBJECTIVES:

1. Describe the effect of negative stories and complaints on organizational culture.
2. Outline the three steps of the author's organizational change cycle.
3. Define the term "tipping point."

You hear the refrain often: "The problem with our organization is our culture. It's why we aren't more innovative, why the wrong people are promoted, why we don't have good leadership." Some of these complaints are justified, if a bit counterproductive.

Workplace learning and performance professionals understand that culture can easily limit much of what we need to do. Because culture is hard to pin down in practical terms, let alone to effectively change for the better, it remains a baffling issue.

However, organizational culture is simpler than our personal cultures, and it is much easier to change than we imagine.

DEFINING ORGANIZATIONAL CULTURE

Organizational culture is different from world cultures, those tapestries of shared histories, languages, beliefs, and foods, which are the source of our identity. Our personal culture affects how we marry, how we raise our children, how we celebrate events, and how we mourn death.

Organizational cultures are not so encompassing, lacking the broad links that help define how we understand ourselves among others. This weakness also implies that organizational cultures are dynamic. The good news is that organizational cultures can adapt and change to new influences quickly.

Organizational cultures are interpretive. Remember when you first took a position in a new company. Remember how strange things seemed, but soon that strangeness seemed to disappear. At that point, you knew the organization's culture so well it didn't seem to exist at all.

For example, bank headquarters are typically grand and luxuriant offices located amid urban centers. They often have bold artwork and distinctive furniture. Whether we acknowledge it or not, these characteristics are purposeful. The company wants you to feel that you are in a place of wealth.

This environment not only influences customers, but also the people who work there. Employees likely will come to espouse this same feeling of wealth and importance.

Most organizations do not rely on such overt references. Instead employees are left on their own to interpret an organization's culture.

IDENTIFYING COMMON REFERENCES

Defining an organization's culture requires being able to identify common organizational references. For example, how do employees describe their colleagues? What are some of the common phrases or stories they tell each other? Such depictions as "bureaucratic" or "people are not valued for their experience and expertise" become a common reference point for interpreting culture whether or not they are accurate.

References become so common in organizations that we often cease to question them. We stop interpreting and simply let the dominant references inform the way we work.

For instance, a co-worker returns from a meeting and says, "As usual, they didn't read any of the documents I sent so the conversation went nowhere. Management doesn't care about the work I'm doing. They were dismissive and wanted to talk about their issues more than all the hard work I did. It's always the same. They just don't care."

If you are listening to this story while preparing for a presentation to a group, you may become hesitant, worried, or defensive. In turn, these attitudes may cause management to be dismissive of your ideas.

Consider another example. You are excited about starting a new position. Many of your new team members welcome you and share your excitement. One person even takes you aside to tell you how things really work. They tell you to avoid John because Susan, the director, really dislikes him.

This co-worker is providing you key references for how you should interpret the organization. The organization tolerates ignorance, and the directors share their personal impressions of people with other employees.

Those are powerful messages to new employees, and ones that will surely influence their interactions from that point forward. This example illustrates that simple stories actually transmit common organizational references.

When stories are negative, reductive, and focused on things that don't work, energy, commitment, innovation, and teamwork suffer. For instance, when you hear such negative stories, do you return to your desk with the energy and commitment? Or do you spend a few hours regrouping, browsing the Internet, and making personal calls? Most of us fall into the latter category. Time we spend regrouping equates to unproductive hours that few organizations can afford to lose.

CHANGING THE CULTURE

Because stories help define an organization's culture, it's easy to use them to change that culture. Simply get people to tell stories that amplify the best aspects of the organization. More important, tell positive stories often to drown out the sound of competing stories.

Typically, organizations try to exemplify their stories by using a common vision and mission statement. Vision provides the aspirations. Mission provides the direction. Unfortunately, vision and mission statements often are poor stories. They either lack drama, or contain too much melodrama. They are abstract and fail to relate to day-to-day roles and responsibilities. They don't engage workers.

Yet, changing an organization's culture does depend on having a common framework. The framework can be used in various ways to get people to share stories about how people across the organization deliver exceptional performance.

Recently, organizations have been developing competency frameworks, which are sets of words and phrases that outline the skills, knowledge, attitudes, and behaviors that the organization respects and that employees need to perform their jobs well.

Competency frameworks not only provide a solid foundation for talent management, but also are well suited for culture change initiatives. They provide a clear backdrop for asking questions and engaging workers to tell stories about what they do well.

Using a competency framework is the best way to generate a number of stories that exemplify the best aspects of the organization and, in the process, to effectively change the culture for the better.

This is an issue of volume. The trick is to counter the negative stories with true stories about positive experiences in the organization without any embellishments or editing.

USING INQUIRY, ENGAGEMENT, & REVIEW

The best way to get people to share good stories is through a cycle of inquiry, engagement, and review.

Inquiry. This includes soliciting answers to questions about how people interpret skill competencies and positive values. You might ask, "Think of someone who exemplifies teamwork. What is it that they do that embodies this competency? How could others learn from this example?" These sorts of questions force people to think differently and invite them to broaden their perspectives regarding organizational values.

Sample tasks in the inquiry phase:

- Conduct a five-question survey that asks people to cite examples of key competencies.
- Conduct a simple survey that invites people to share what they value about working in the organization.
- Set up a peer-interview process, whereby two people are given a questionnaire and asked to interview each other. Post interview results in a common forum.

Engagement. This builds on common themes identified during the inquiry phase by asking other people to comment on the stories that were shared. You might say, "Seventy-three percent of the people surveyed said that the best collaboration in our organization happens among small, informal groups that share a passion for a particular subject. Can you cite any examples of this type of collaboration that you've experienced?"

Sample tasks in the engagement phase include

- Conduct a survey that invites people to match specific workplace challenges with the competencies or other common references. Ask them to describe how they exemplified specific competencies to meet the challenge.
- Conduct a debate among members of the senior management team. For example, have management debate which of the competencies is the most important given the organization's mandate.
- Invite general staff to describe why specific competencies are important and how their managers exemplify those competencies.
- Introduce training and development activities that align with the organization's competencies.

Review. This action strives to uncover the best stories from the engagement and inquiry phases, as well as determine how best to circulate these stories throughout the organization. It also requires some investigation of patterns and trends in how people relate to the common references, competencies, or other frameworks that extol the organization's best performance and values.

In particular, you want to identify common phrases, similarly stated challenges, or a typical story about high-performing individuals. Circulate common stories as broadly as possible, either via newsletters, the intranet, or on bulletin boards in break rooms.

In addition, when you spot a trend in the review phase, be sure to highlight it in the next cycle of inquiry, engagement, and review. For instance, if multiple employees report, "Our organization has some of the brightest minds in the field," your next cycle of inquiry questions could include, "How does the fact that the organization has some of the brightest minds in the field enable it to build partnerships?"

SPOTTING THE TIPPING POINT

How do you know how many times to repeat the inquiry, engagement, and review cycle? This is difficult to determine, but you'll likely know it when you get there. Once you reach that point, changing the culture will continue on its own.

Organizations, like all systems, experience tipping points—points where system inputs are sufficient enough to cause exponential changes in a new direction. For example, physicians use this concept when prescribing medicines. They know precisely how much medicine will be needed to cause sufficient change to the system to combat bacteria or germs.

Unfortunately, we don't have that level of scientific acumen in organizational dynamics. Instead, we need to rely on a keen eye and investigative talent to spot common cultural indicators.

Architects of cultural change programs must be patient and trust the process. It will be extremely difficult at times to see any change and to listen to pessimistic stories that disrupt work and negatively influence the organization's culture. In fact, negative stories are sometimes told more often during the process, only to go silent after a short while.

This organizational dynamic is difficult to track. However, if you cease the process, negative stories will quickly overwhelm any good you may have started. Again, it's like medicine. If you stop taking tablets before the prescription runs out, you risk having the infection return quickly and with full force.

USING YOUR SYSTEMS

Once a tipping point has occurred and you are satisfied that the organization is adopting a positive culture, you need to ensure that all of your systems, such as recruitment, training, talent management, and performance management, reflect and champion it. You want the new, positive stories to become so common that people can't remember what came before.

If you follow the inquiry, engagement, and review process, you will undoubtedly create culture change. More importantly, you will definitely be surprised at how effective, productive, content, and committed employees become, and how much better it is to work at your organization.

ABOUT THE AUTHOR

Dorian LaGuardia is a Europe-based consultant; dorian.laguardia@thirdreef.org.

From: Dorian LaGuardia, "Organizational Culture," *T+D* 62, no. 3 (March 2008): 56-61. Used wtih permission.

15.2 Manage Change – Not the Chaos Caused by Change

By Beverly Goldberg

OBJECTIVES:

4. Define the three stages of the change communication framework.
5. Identify the major roadblock to managing change successfully.
6. List necessary steps for successfully managing change in an organization.

In the 1400s, in the city of Mainz, Germany, Johannes Gutenberg invented movable type and revolutionized the world. Typesetting, letter by letter, was done by hand from then until the 1880s, when mechanized typesetting – linotype – made it possible to se whole lines of type in a single operation. In the 1950s, cold type became commonplace, allowing for even faster typesetting; then, in the mid-to-late '80s, the development of affordable computers that enabled almost instantaneous typesetting began a new publishing revolution that has not ended. Today, desktop publishing means quick, easy and less costly delivery of an incredible variety of publications.

The times between these changes – 480 years, 70 years and 20 years – are a good indicator of the increased speed with which change is taking place. As a result, change must be managed differently: The classic change management techniques that helped organization institutionalize change are no longer adequate to meet today's needs.

According to classic theory, change management required several steps: unfreezing the organization's existing culture so that a change could be brought in, creating cognitive recognition to open the workforce to what was new, and then refreezing the culture once the change was accepted. The idea was that the culture would then remain constant until the next change came along. That may have worked when change came about only every 20 years or so. In a world in which change seems to occur every 20 minutes, a new framework for managing change is necessary (see chart below).

In this framework, the first or Static Stage, as in classic change management theory, calls for unfreezing the current culture by convincing employees that the organization is changing and the changes have the strong support of senior management. The second or Fluid Stage begins when employees start to understand that the changes will benefit them as well as the organization. They recognize the whys and wherefores of what is new and they accept it. Then, breaking with the classic idea of refreezing the culture as a final stage, the culture is moved to a Dynamic Stage, where people work with the new machines or processes and act in the new manner, but await – and even anticipate – the next changes that will be made. In other words, openness to change and anticipating change become the mind-set of the organization.

A LABORIOUS TASK

Helping employees through change is not an easy task. It requires formal programs that must be introduced gradually and managed with care and thought. They must be planned for with the same care as the new strategy or technology that is making change necessary. Each part of a change program must be constructed so that the need to be prepared for constant change reaches – and is understood by – employees at all levels. But this is far easier said than done.

The major roadblock to managing change successfully is the fact that change does not happen in isolation. Take the case of a major East Coast insurance firm that encountered more than a few pitfalls when it attempted to bring about change.

The technology services group was trying to achieve the goal of employee acceptance of computer-aided software engineering (CASE), a leading-edge technology for developing computer programs that requires a number of changes in the way people work and the way they think about their work. CASE shifts the emphasis in developing programs from writing the programs, which is the job of programmers, to analyzing the business function the pro

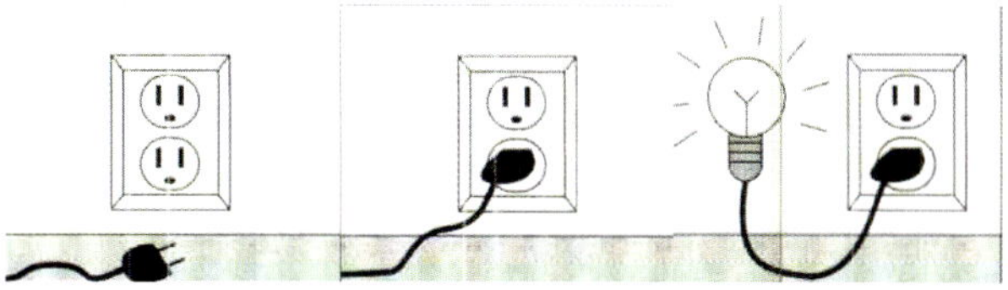
Static
Fluid
Dynamic
The mind-set of the organization is frozen and needs:
Senior-level determination to take a new road
Employee awareness that survival means change
As the mind-set opens, build recognition, understanding, and knowledge by:
Demonstrations
Opportunities for learning
Training
The mind-set is open to what is new and other projects:
Anticipatory capabilities
Flexibility
Acceptance of continuous change and learning

gram will serve and then having the computer write the program instead.

Before CASE, business groups had to explain to a programmer that they wanted information regarding the health risks of smoking, for example. The programmer then wrote a program that pulled the necessary information from the company's computers. Once the basic program was written, those who requested the information would review it. Inevitably, the program would require changes so that it would retrieve the specific information needed, for example, collect information on health risks of smoking by males in a certain income bracket.

With CASE, businesspeople spend a great deal of time analyzing the business needs that the program will serve. The analysis is then fed into a computer on which CASE has been installed and the program is generated automatically. The detailed upfront business analysis prevents many of the problems that develop when a programmer designs a program without a clear understanding of the purpose of the program.

A number of pilot projects were underway at the company. Selected groups were developing a number of major computer applications in this new way. CASE was the subject of a great deal of speculation by the technology services group because switching to CASE technology created fears among programmers about the need for their skills in the future. At the same time, the businesspeople were concerned about the amount of time they were spending doing upfront analysis instead of concentrating on their jobs. These two groups also had trouble communicating because of their different backgrounds and "special" terminology, acronyms and jargon commonly used only within their own groups. Also, they had never worked together before.

MACHIAVELLIAN CHALLENGE

It should have been easy to explain to the programmers that their skills would be used differently; they could do much of the upfront analysis if they enhanced and updated their skills. The businesspeople needed to believe that the time they invested upfront would pay back in programs that would allow them to do their jobs better. This turned out to be far from easy.

Changes of any sort are never easy. Machiavelli's descriptions of the problem faced by those who must bring about change is just as apt today as it was when he wrote it in the 15th century: "It must be remembered that there is nothing more difficult to plan, more doubtful of success, nor more dangerous to manage, than the creation of a new system. For the initiator has the enmity of all who would profit by the preservation of the institutions and merely lukewarm defenders in those who would gain by the new one."

In this case, the difficulties normally associated with bringing change were magnified by an organizational history of upheaval, rumors about the pilot programs, the launching of new initiatives, the cursory attempts to explain what was going on without addressing the specific change, and a basic distrust of the abilities of other departments, especially communications and human resources, to help facilitate the change.

TRUST IN MANAGEMENT

The employees in the technology division did not believe that senior management was telling them the truth because of a recent history of reorganizations, large workforce reductions and the belief that further reorganizations were likely. Moreover, before the most recent reorganization, the company had denied rumors that reductions in workforce would occur until hours before the announcement was made. The combination of historical evidence and a lack of trust made every statement issued by senior management suspect.

Senior management recognized the need to do something to stop the negative speculation about CASE fairly late in the game. The pilot projects had been underway for almost a year before management became aware that, instead of greeting CASE with enthusiasm, employees were apathetic, secretive and reluctant to go off-site to attend training sessions.

Coincidental with the introduction of CASE, a number of systems, including e-mail, were slowly being put in place to make the office function more efficiently. In addition, another group within the division was creating an organization-wide data bank that would allow senior management to access financial information in real time. Unfortunately, the data bank could not be accessed by the hardware and software used for the new office system. Confusion over what was happening and frustration in the face of supposedly "new and improved" technology that actually was wasting time and energy created strong pockets of resistance to change.

The few attempts to help employees accept the changes taking place were based on textbook change techniques, and thus ignored the problems specific to the organization. For example, given the organization's history of upheaval, employees were not satisfied hearing middle managers say, "Word is that the head of the company supports this." Nor were they satisfied by a letter from the CEO saying, "We are going to move into the future rapidly."

Management in this department was unhappy with the lack of attention its work was given in the company newsletter and the communications vehicles tailored to the department. It believed the communications people did not understand the tools or the language of technology, and that the human resources department was ineffective. In fact, human resources frequently sent applicants who had the wrong skills for the job. On the other hand, the technology department's managers never explained to human resources the specific skills they were seeking. Instead, they simply rejected the department as worthless, never finding out, for example, that there were video training tapes available that discussed problems that occur when change takes place.

TIPS for PRACTICAL APPLICATION

To manage change successfully, each step must be communicated to employees using various communications tools, such as paper or voice mail memos, bulletins, newsletters, focus groups, forums, brainstorming sessions, meetings, training tapes, multimedia presentations, performance rewards, classes, and broad circulation of specific information, particularly articles. Each of the items on the checklist below represents a necessary step in managing change:

- Stop the rumor mill.
- Begin a strong, targeted communications campaign.
- Make senior management commitment clear.
- Make employees aware of why the change is necessary.
- Achieve buy-in at all levels.
- Break down the barriers between employees.
- Provide training.
- Ensure that anticipatory capability is built into the culture.

CHANGE AS WARFARE

Since there were so many landmines planted and the change was already underway, managing this change required a more intensive, longer program than usual. It took three to four months to establish enough momentum so that it could continue on its own (normally, one to two months is sufficient to get a program up and running). The first stumbling block was the inability to pull together the ideal team – one or two members of the group driving the change, as well as members of the communications and human resources groups – because of the technology group's negative feelings about those two groups.

Although the client company assigned a very strong, senior-level individual to work on the change program, one person is not the same as a team. First, having a trained team is an added value for a company because it enables them to develop their next change program on their own.

Moreover, since change is continuous, the team also can be the beginning step of a long-term effort to build a learning environment. In addition, if one member of a team leaves, the organization does not lose all of the skills. Furthermore, the mandate not to use the communications department meant that a lot of time was spent building communications vehicles from scratch. For example, the major communications vehicle, a new newsletter directed at the change to CASE, had to be designed and produced without the use of internal resources.

The final step before putting a change program in place was to "take the temperature" of the organization more thoroughly, checking that the problems described by management were real and that the ones they were unaware of were taken into account. The department had more than 1,000 employees and a set of interview and focus group sessions were needed to ensure that the right techniques were selected. Following are the five major components of the program:

- *Opening channels of communications.* Since what people don't know can hurt the organization ,it is important to open a dialogue between management and employees to forestall speculation. Knowledge prevents misinformation and disinformation and reduces the chance that playing politics against change will succeed. Moreover, letting people know what lies ahead helps them face the uncertainty brought by change and eases their adjustment to the new reality. It starts them on the road to understanding.

The major method of communicating to a group this size was the newsletter. The first edition contained an open letter to that division from the head of the organization. It clearly stated his determination to make the shift to CASE technology. The newsletter also provided information about other organizations that had successfully introduced this technology and information about CASE itself. Subsequent editions continued to drive home the message of upper management support and provide examples of successes.

- *Creating visionaries and change agents.* The conflict between those who believe that the best way to drive change through the organization is from the top down and those who believe change must be driven across by example is extremely easy to resolve. Change is best and most firmly driven deep into the organization when management takes a two-pronged approach, establishing visionaries at the senior level who drive change down into the organization, and change agents or champions at the middle management level who drive it across the organization. In other words, a double effort is needed to ensure successful buy-in at all levels and across all functional areas.

- *Developing a learning environment.* Continual learning and an educational environment help employees at all levels and functional areas not only to accept the fact that change is critical to success, but also to search for future change opportunities. Employees learn that anticipating the changes being made by competitors is critical to the company's survival. And building understanding of the continuing nature of change brings greater acceptance of current change.

- *Providing training.* Employees were informed that they would be given every opportunity to learn the new skills that would enable them to work with CASE. The initial training offered was not, however, in the form of training classes during business hours. A "Choice Learning Lab" approach was decided upon, one that allowed interested employees to come in early or stay late to explore new technology. Those who showed interest were invited to take formal classes. The result of this approach is that the first to learn are the most interested and confident of their abilities, and are most likely to become change agents enthusiastically promoting what they learn.
- *Establishing a team approach.* Getting people who had previously had little contact with one another to work together comfortably and to understand one another was critical to the successful implementation of CASE. A "Book of Words," a dictionary containing definitions of the acronyms and jargon used by businesspeople and technologists, was created to help communication and understanding.

In addition, "Partnership in Action" sessions were held with members of the pilot project discussing the problems they had in the beginning with groups that were entering training.

Bringing change to an organization is a difficult process. It requires the right mix of psychological, communications and management skills –and a great deal of empathy and understanding of the pain it causes people. In many ways it is like parenting: You know you are doing it for the child's own good, but it still hurts to watch the growing pains.

The rewards, however, can be great. A successful change program opens the organizations to the specific change needed at the moment and, if handled correctly, creates a dynamic environment that will enable the organization to remain competitive in the future. Thus, it helps individuals, organizations – and the American economy. And that, of course, helps us all.

ABOUT THE AUTHOR

Beverly Goldberg, vice president and director of publications of the Twentieth Century Fund, a New York City-based think tank, is also a partner in Siberg Associates Inc., management consultants.

From: Beverly Goldberg, "Manage Change – Not the Chaos Caused by Change," *Management Review* 81, no. 11 (November 1992): 39-45.

15.3 Keeping Change on Track

By Richard Bevan

OBJECTIVES:

7. Name several pitfalls that can cause change to get off track.
8. State and define seven core factors in successful change management
9. Identify three key aspects of organizations that readily adapt to change.

It's rare to find a business leader who is not involved with planning or managing a change process of some kind. Whether major (a plant shutdown, merger, installation of a new IT system) or on a smaller-scale (engagement of a new leader, sales force reorganization, new compensation plans), change can raise questions and concerns, present operational challenges, and impose demands on time and resources. The cost of managing the process may increase greatly if attention is diverted from day-to-day work with customers and prospects.

We often see significant changes implemented with scant attention to identifying and addressing the challenges it may create among those involved, the questions it will raise, and the issues and needs it will generate. As a result, many change efforts stumble, and some fail entirely. For example, Peter Senge states that two-thirds or more of total quality management (TQM) programs and reengineering initiatives fail.1 John Kotter noted that few of the companies he studied were successful in making major changes to their ways of doing business.2

A simple series of questions will enable you to assess quickly where the process is going well and where it might need strengthening. These questions assess the status of the change initiative in relation to a set of seven core factors typically present in successful change.

Before reviewing those factors and the questions we will briefly consider the nature of change within organizations and the reason that—more often than not—the process doesn't run entirely smoothly.

WHY IS CHANGE SO OFTEN A CHALLENGE?

The characteristics and processes that predict effective adaptation to change have been defined and described by many leaders, researchers, writers, and educators. At its most basic, effective management of change requires leaders to be very clear about the purpose and the process; seek input and information from those involved and affected; deploy sufficient resources to manage the transition without losing focus on day-to-day business processes; and maintain an effective multi-directional flow of communication and information.

The attributes are straightforward, readily implemented, and perhaps considered self-evident. Most people manage change continually: at home, in recreation and volunteer activity, and at work. They have an intuitive understanding of what needs to happen if change is to move forward. Even if they haven't consciously thought about or documented the principles, they do what makes sense. They consult people, discuss the alternatives, try to anticipate and plan around the obstacles, adapt their plans as needed, get on with it, and address issues and challenges along the way.

Yet when organizations implement change, these straightforward needs are often missed. The intent and the broad strategy get the attention; the details of execution are forgotten. We repeatedly see change initiatives within organizations stumble and sometimes fail entirely. We see participants in the process who are unclear about the purpose, the impact, and their role. We see insufficient resources assigned to much-needed systems changes or to prepare or train people for new roles and behaviors. We see managers and supervisors unable to respond to questions and concerns from their teams.

If the core needs are well known—and perhaps even intuitive—why does change within the organization so often present such challenges and run into difficulties? Part of the reason is that leaders and sponsors of change initiatives already face a heavy workload and multiple initiatives and activities. They rely on the so-called "memo and conference call approach" and assign far too few resources to managing the process of transition.

Significant change calls for proportionately significant effort and resources: for planning, communicating, training, and many other activities. In some cases, and especially where there is resistance to the change, the process of persuasion, engagement, and adoption calls for continuing time and commitment from managers at all levels. Initiating change will likely create a complex and extensive set of issues, questions, and unresolved concerns. These, in turn, create the need for a continuing and demanding process of explaining, discussing, persuading, and planning.

Many years of working with change provided opportunities to generate a substantial list of the potential pitfalls. We have also seen the tactics, strategies, and processes that provide a positive effect and enable change processes to succeed. These inputs suggest a framework of characteristics or attributes that can guide successful management of change. Of course, no single element in this framework—or even several of them—can guarantee a successful outcome. The absence of even one will certainly lead to difficulties, and absence of more than one will cause the challenges to grow exponentially.

For example, change rarely succeeds without a clear awareness of purpose and process. People need to understand why the change is needed, how it will be accomplished, their role in the process, and what it means for them at an individual level.

CASE HISTORY: ANTICIPATE THE ISSUES AND PLAN THE RESPONSE

A medical-equipment company was expanding quickly by internal growth as well as through acquisitions. The announcement about consolidating into a single new Midwest plant (from several locations in the United States and Mexico) raised many questions and concerns across the organization.

Would people lose their jobs or face relocation? How would the organization retain expertise and apply it in the new plant? Was the change motivated by cost reduction and, if so, what other approaches were explored? What impact would this have on production cycles, product lines, and development teams? How would this affect pay and other conditions?

A transition steering group was asked to develop the implementation plan. As the leader commented, "There wasn't much source information for us. The board made the decision based on a presentation from the planning team. They talked about industry economics and financial implications but didn't get into the implications for the organization and our people—how we would actually get this done. And the rationale hadn't been summarized in a clear and concise way."

The team interviewed key executives for brief outlines of purpose, rationale, process, and implementation steps. They conducted research among employees and other stakeholders (including customers) to explore and understand their reactions to the planned changes and the implications.

The resulting database of questions, issues, and ideas provided the raw material for developing responses and action plans. The implementation plan was built on this foundation, including activities to address training needs, manager support, alignment of reward systems, communication programs and processes, and many other actions in support of the change.

The research jump-started planning, identified major issues to address, and clarified priorities. It provided a clear starting point and foundation for implementation planning. Follow-up research in specific areas and functions evolved into a key change-management tool.

SOME OF THE PITFALLS

Change is often imposed without advance assessment of the issues, questions, concerns, and ideas of the stakeholders—those most involved and most able to influence the outcome. Yet if questions remain unanswered and concerns unaddressed, employees may be distracted and distressed. This carries a high potential cost. How often have we experienced the frustration of dealing with a distracted employee in a business providing a product or service? Very often, a poorly managed change process lies behind that negative customer experience.

Even if change efforts are well planned and executed, those involved and affected will ask questions and analyze purpose and implications. This is especially true if they haven't been involved in the planning. "They never consulted me," is a common refrain, and it will slow down the process and create challenges and additional workload for line managers. Acknowledging the level and nature of concerns and questions and building a degree of involvement can also provide insight and information about how best to manage the process. The energy and focus of employees, frequently encountered as a challenge or obstacle, can be harnessed and turned into a powerful tool for change.

Following are a few of the pitfalls that cause change to get off track:

- *Ineffective or missing business case:* Managers communicate a case for change that is unrealistic or incomplete; it isn't readily understood. Here is one of many areas where a robust, concise business case document is invaluable.

- *Costs not recognized:* The costs of implementing and supporting change are not planned for or adequately acknowledged. Resources need to be allocated, workloads adjusted, and responsibilities reassigned.

- *Systems not aligned:* Existing processes and systems (e.g., rewards, training, and information) don't support the new model. Change ripples across many areas and functions, and these need to be aligned with the new direction.

- *Limited and one-directional communication:* Leaders expect to persuade and inform by one-way communication. Audiences have limited opportunity to ask questions, offer ideas, or engage in discussion about the changes.

- *Line management support not built:* Line managers don't support the direction and approach. They share the uncertainty and concerns of employees and need to be informed, persuaded, and engaged.

- *Lack of insight into stakeholder issues:* Leaders assume they know what people think. They fail to identify key concerns and obstacles. They need to listen informally, and at an institutional level, to the questions and concerns of stakeholders and (most importantly) to their ideas.

- *Minimal involvement:* Input, questions, and ideas aren't recorded and documented. As a result, responses and tactics don't reflect the needs. A continually revised FAQ document, available online, can be of great value.

- *Success assumed*: Leaders make a premature assumption of success and fail to follow up, support, and drive continuing change. Awareness of these and other pitfalls doesn't ensure success, but it does provide ideas about where change most often gets off track and opportunities to implement course corrections. Each pitfall has a positive counterpart — a proactive measure to support and facilitate change.

CORE FACTORS IN SUCCESSFUL CHANGE MANAGEMENT

These seven factors summarize the conditions, resources, and processes that support successful change.

- *Clarity.* Be clear and unambiguous about the purpose of the change, its direction, and the approach.

- *Engagement.* Build a sense of ownership, belonging, and commitment; consult with and involve the people who will be affected by the change.

- *Resources.* Put the needed resources in place (e.g., financial, human, and technical) to enable the change.

- *Alignment.* Ensure that systems and processes (e.g., rewards, information, accounting, and training) support the change.

- *Leadership.* Guide, train, and equip leaders at every level so that they display consistent commitment to the change.

- *Communication.* Facilitate an effective two-way flow of information; be aware of issues and questions; provide timely responses.

- *Tracking:* Establish clear goals; assess progress against these; adjust and fine-tune as necessary.

The set of factors listed here aligns with models and frameworks developed and applied by many writers, educators, and leaders. These include John Kotter2 and Daryl Conner,3 as well as Kurt Lewin,4 one of the earliest commentators on change and still fully relevant. The challenge doesn't lie in understanding the process, or even in putting together an effective plan: it's in putting the plan into action and sustaining the effort.

MAKING IT HAPPEN

"Everyone knows you have to do these things" is a frequent comment when the elements of successful change management are outlined. When we ask if those elements have been put into practice, it often turns out that perhaps some of the details were overlooked; maybe most of the details; sometimes, all of them. In many change initiatives, large-scale and small, at least one of the core principles (and, typically, several of them) is not followed.

Senior leaders often say of their employees, "They're smart; they'll figure it out." Yes, they are indeed smart. They figure out that the direction isn't clear and the planning is imperfect. They discover that those leading the initiative, already committed to a heavy workload, have little time to focus on the new task. They figure out that they need a great deal more convincing that this is a change that warrants their involvement.

The details are what make change work for those whom it impacts most sharply. It's hard work to make a significant additional effort while continuing to run a complex business, but there's a high price if that effort is not sustained. Employees get distracted and unmotivated; customers' needs get ignored or forgotten; and questions, issues, concerns, and distractions consume managers.

Change can happen without all seven core factors in place, but it's likely to be difficult, expensive, and painful — for your customers as well as your employees.

A SIMPLE ASSESSMENT FRAMEWORK

The questions below can be asked and answered informally, in a series of conversations or discussions with those involved, or more formally—for example, through an online survey of stakeholders. They provide a means of quickly assessing status and key issues, with the negative items offering a guide to where attention is needed to keep the process on track.

At the least, the areas of concern offer direction for additional research and study. The data can serve as a primary driver of planning. The issues and questions you uncover in the research process will determine the activities (including communication, process support, and training) that form your tactics for redirection.

Following are the assessment questions — two for each factor, one primary and one secondary or follow-up — together with some examples of tactics that support positive outcomes.

- *Clarity.* Are the purpose, direction, and approach defined and documented clearly? Are these understood and accepted by key stakeholder groups?

- *Engagement.* Have you engaged individuals and groups who can influence the outcome by involving them in the process? If so, have you acknowledged their input and ideas and applied them to planning and action?

- *Resources.* Are needed resources (e.g., financial, human, and technical) in place and available? Is a strong and effective team ready to lead and guide the change process?

- *Alignment.* Do systems and processes (e.g., rewards, information, accounting, communication, and training) support the change? Have you identified, developed, and implemented needed changes to these systems?

- *Leadership.* Are leaders at all levels of the organization involved in and committed to the change? Do leaders and managers follow up on issues, provide guidance and support, and proactively manage the process?

- *Communication.* Is clear, timely, and complete information available to stakeholders involved in and/or affected by the change? Do these groups and individuals have access to information and a way of providing input and feedback?

- *Tracking.* Are systems in place to assess progress and identify issues to address? Are adjustments implemented as necessary and is information continuing to flow?

TACTICS FOR SUPPORTING THE CHANGE PROCESS

The responses to these questions will suggest areas for action and perhaps offer ideas for some of the actions themselves. Ideally, ask the questions in a manner or setting that permits responses that go beyond a simple answer to the question. For example, if they're raised in a discussion led by a team leader, there is an opportunity to ask team members what ideas they have for facilitating, supporting, and driving the change. The result is that awareness of the change process status is complemented with ideas for addressing issues and correcting issues and problems.

Every change is different, but some consistent themes emerge. Following are just a few examples, relating to two of the seven factors, of actions that can support and refresh the change effort.

Clarity

- Develop and distribute a summary document to drive clarity and serve as a reference source on the purpose and process of change.

- Distribute the summary. Use it as a platform on which to build all communication (internal and external) related to the change.

- Create a brief elevator pitch for managers— what's changing and how the transition will be accomplished.

- Develop other tools to assist in the process; for example, a brief PowerPoint® deck for executives and others to use in discussing the changes with their teams.

- Provide managers with talking points and suggested responses to key questions.

- Maintain and manage the summary. Seek input and comment; keep it current, accurate, and complete.

- Provide online access to the current version and enable input, questions, and discussion.

Leadership

- Ensure that the primary sponsor(s) of the change (in some cases, and certainly for major changes such as mergers or acquisitions, this may be the chief executive) is visible, accessible, and driving the process consistently.

- Engage leaders at other levels in planning and implementation; provide guidance and training as needed.

- Encourage leaders to promote behaviors and actions that will support the change.

- Implement processes and activities to maintain leadership engagement—for example, weekly conference calls, regular e-mail bulletins, online forums, blogs and other interactive media, and planning-review meetings.

- Address concerns that leaders may have about their roles and responsibilities in and after the change process; provide guidance, tools, and support.

MANAGE CHANGE BEFORE IT HAPPENS

Many strategies for managing change are not proactive; they focus on the defined period when change is actually occurring. These include the phase following an acquisition, process redesign, division-wide reorganization, or the response to a competitive threat. Such change management efforts often fall short of expectations in terms of accomplishment and employee satisfaction. You can't always create the core factors in successful change at short notice: They need to be in place.

Change comes more readily to the organization that has:

- A clear mission and strategy that guides and informs the goals of teams and individuals.
- Supportive leaders at every level who effectively engage, motivate, and communicate with their teams.
- Employees who are engaged, informed, and involved.

Creating change readiness means managing in a way that encourages engagement, commitment, aspiration, and adaptability. A transition is far more likely to achieve success when the leadership style, work climate, and environment are already receptive to change.

References
1. Peter Senge, *The Dance of Change*, Doubleday, 1999.
2. John Kotter, *Leading Change*, Harvard Business School Press, 1996.
3. Daryl Conner, *Managing at the Speed of Change*, Random House, 1992.
4. Kurt Lewin. *Resolving Social Conflicts: Selected Papers on Group Dynamics*, Harper & Row, 1948.

ABOUT THE AUTHOR

After early experience in manufacturing management, Richard Bevan worked for Towers Perrin (now Towers Watson) in Europe, Australia, and the United States, including five years leading the firm's worldwide communication consulting practice. In 1995 he started his own firm, C2K Consulting. He was an external faculty member for the University of Washington Executive MBA program where he developed and taught a course in managing change. He currently serves on the board of advisors for ROI Communication and as strategic communication adviser for Elliott Avenue Associates, This article draws on material from his new book Changemaking.

From: Richard Bevan, "Keeping Change on Track," *The Journal For Quality & Participation* 34, no. 1 (April 2011): 4-9. Used with permission.

15.4 Developing an Innovative Culture

By Erika Agin and Tracy Gibson

OBJECTIVES:

10. List ways that leaders can encourage innovation in their organization.
11. Describe the benefits of open communication.
12. Discuss the role of motivation in developing a creative workforce.

Leadership development has evolved with the times. Today, engaging a workforce and grooming young employees for future leadership positions requires a focus on innovation, creativity, and open communication.

As members of the young and energetic workforce, we are where leaders need to turn for innovation. The entry-level employee to mid-level manager has a need to do work that is meaningful, or at the very least have a say in how it could be done. Nothing provides more inspiration for creative ideas than the opportunity to make a positive difference in work processes or outcomes and to be rewarded and acknowledged for it.

YOUNG EMPLOYEES UNLEASH INNOVATION

Leaders of organizations who desire innovation in their business cultures should ask for input in decision making and guide their staff members through creative thinking processes to ensure followers are on the right track. In addition to stimulating innovative ideas, allowing employees at all levels to take part in the decision-making process will facilitate transformational leadership development for the future of the organization.

When leaders give followers the freedom to make decisions, they are enabling employees to experiment with ideas in a safe environment and challenge themselves with a new way of thinking. With the ability to add to the work process, employees will begin interacting in a way that supports innovative ideas and influences the future of the business.

OPEN COMMUNICATION INSPIRES CREATIVITY

An imperative aspect of innovation is companywide communication that generates trust and encourages information exchange. A culture that allows communication to flow openly and evenly across all levels and departments will find that employees even in entry-level positions have the tools necessary to envision opportunities for better ideas.

Most organizations do not listen to the ideas of those lower in the ranks, but those employees are often bright students right out of college or the best of the best who transferred from other organizations. The fresh ideas produced from the newest additions have the potential to improve the company in ways never thought possible by a seasoned workforce. An employee that is kept out of the chain of information exchange will be less motivated and will suffer from diminished levels of creativity. It is important that everyone in the organization is included in communications. The mission and vision of an organization sets the tone for the culture that eventually develops. The mission and vision should be communicated daily and should foster openness in communication in all directions. Many organizations only give lip service to open communication or open door policies. As a result, a lack of trust and dismal levels of creativity develop within the culture.

The premise that leadership has all of the answers and that the followers should not question is a stale and bureaucratic ideology that is not conducive to inspiring innovative thinking. Organizations that want to become more innovative need a mission and vision that encourage ideas from their workforce and actively seek input from all departments and across all levels, ultimately empowering their people.

EMPOWERING EMPLOYEES TOWARD NEW IDEAS

When a workforce is empowered to make decisions, employees are then able to think outside of the box and drive the organization in new and innovative ways. The transfer of power should take place once the employee is fully integrated and capable of making sound decisions. The authority to make decisions should be offered in a progressive manner, so the employee maintains her creative momentum.

A wealth of potential is unlocked when a workforce is empowered to do the work the best way it sees fit. The momentum generated from empowered people can set a fully committed organization on the course of ongoing improvement. Empowerment needs to become a part of the culture and should begin as soon as the employee feels comfortable making decisions.

Since its people are a company's biggest asset, it is important that employee ideas are rewarded by implementing those ideas to improve work processes and outcomes. After all, they do the work every day, so who better to ask about doing it a better way?

By validating forward-thinking ideas, the organization will provide inspiration to the rest of the workforce to continuously look for ways to cut costs, save time, and produce more. If the employees do not think their ideas are welcome, then they will never disclose them. The culture should celebrate and embrace smart business ideas no matter from whom they come in the organization.

MOVING NEW THINKING THROUGHOUT THE ORGANIZATION

Quality employees who can provide an organization with innovation will require options for individual development to impact the cultural change toward innovation. It is to the company's advantage to use its younger employees' needs for skill development and job changing to propel innovation throughout the culture.

An employee moving from one department to another will be able to transfer new ways of thinking from one place and use them in the new role. The steady move of employees from one sector to the next will take innovation to a whole new level and decrease the time it takes to instill innovation throughout the entire organization. If the less experienced— and in most cases younger— employees within the organization feel stagnant or stifled for too long, they will be less likely to remain engaged.

An excellent solution to appease this appetite for knowledge is to ensure that leadership is dedicated to assisting followers with learning new job roles and thinking in new and innovative ways about their current positions. When changing departments and titles, the employees will drive innovation forward and improve work processes with a deeper understanding of the bigger organizational picture.

THE SUPERVISOR'S ROLE IN DRIVING INNOVATION

Younger workers in organizations are not influenced by titles but instead by the leader's capabilities and willingness to share knowledge. Ideal leaders will set an example of versatility and provide a safe environment where trust and candor are highly valued. Successful assimilation of innovation into the organizational culture requires leadership to foster and develop innovation among their followers.

The actions and behaviors of midlevel managers are directly related to the creativity that an organization will produce. With in-depth leadership training, proper accountability, and daily communication about leadership's responsibilities to foster creativity and trust, mid-level managers can rise to the standards necessary to inspire innovation and grow the next generation of innovative leaders.

Leadership must understand the importance of truly knowing their followers and what motivates them if they want to inspire them to contribute creative ideas. When a supervisor focuses an employee in areas that are naturally motivating for him, the employee has the potential to far exceed average performance.

Each person is motivated by different things, and it is the direct supervisor's job to get to know employees well enough to align their interests with overall job goals. A motivated employee will find that she is constantly pushing herself to improve the work process and outcomes generated. A supervisor who asks questions and targets specific jobs that match the employee's innate motivations will engender a staff that searches for new ways of thinking about what they do best.

Management can propel the organization in new and innovative ways and inspire a workforce to meet the company's vision of innovation by validating innovative behavior each day in ways that appeal to employees. One of a leader's responsibilities must include knowing what followers deem to be a reward or a punishment. If a particular employee finds leaving work early to be a reward, then leadership could leverage that fact to congratulate the employee for a job well done.

Job satisfaction plays a major role in stimulating new and productive ideas. Employees must be given the opportunity to do what they find to be satisfying in their jobs to harness the creativity necessary to establish an innovative thought process.

In the diversity of the modern work environment, many new and exciting ideas can be generated and are a direct reflection of the culture of the organization. Choosing the right person for the right job will create a comfort level that encourages invention. If workers predominately occupy jobs that are satisfying for them, the organization will see much success in the output and quality of ideas.

CONCLUDING THOUGHTS

In an ever-changing business world, innovation is the key to launching business organizations into the future. Although many cultural qualities are necessary to construct the proper environment for innovation, communication is by far the most important. By encouraging open communication and taking a genuine interest in the development of all levels of the organization, employees will have the drive and inspiration to develop fresh and groundbreaking ideas.

Transformational leadership coupled with a company-wide commitment to employee development provides the means to inspire a workforce toward productive invention. If we are to develop more innovative cultures in our organizations, we will need to adopt more committed and less controlling behavior to empower creativity in the workforce.

ABOUT THE AUTHORS

Erika Agin is an organizational leadership and human resource management professional based at Walt Disney World in Orlando, Florida, where she researches employee benefits that contribute to overall health and well-being; erika.agin@yahoo. com.

Tracy Gibson is an assistant professor of organizational leadership at Chapman University in Silverdale, Washington; tgibson@chapman.edu.

From: Erika Agin and Tracy Gibson, "Developing an Innovative Culture," *T+D* 64, no. 7 (July 2010): 52-55.

15.5 The Twenty-First Century Leader: Social Artist, Spiritual Visionary, & Cultural Innovator

By Fahri Karakas

OBJECTIVES:

13. Describe each of the three creative roles that twenty-first century leaders perform.
14. Define the term "mind-set."
15. Define the term "heart-set."

The old leadership model is giving way to a new twenty-first-century paradigm for navigating in an age of uncertainty, complexity, interdependency, globalization, and accelerating change. Drawing from chaos theory, nonlinear dynamics, quantum mechanics, and other disciplines, the author sees the emergence of three new creative roles for leaders— social artist, spiritual visionary, and cultural innovator—which call for a new leadership mindset, and for tapping the powers of the heart. A new holistic skill set with eleven domains encompasses the values, perspectives, and competencies needed to lead organizations and employees to their full potential in the new century.

As we stand seven years into the twenty-first century, one thing is abundantly clear: We aren't in the twentieth century anymore. Postmodern terror, global warming, the rapid proliferation of technology, globalization, hyper-competition for resources and markets, corporate mega-scandals, diversity in markets and the labor force, the widening gap between developed and undeveloped regions— these are just a few signposts of the new age. Although we cannot forecast what the second half of the century will look like—any more than someone in 1907 could have accurately predicted the shape of civilization in 1957—the trends that will characterize the next several decades are here already, disrupting the old order and posing unprecedented challenges, particularly for societies and organizations whose leaders are ill equipped to deal with the new order.

A PARADIGM SHIFT FOR LEADERSHIP

How does the new world for society and business differ from the old? The new world is complex, dynamic, fast paced, and knowledge intensive. Markets, products, and organizations have become global, diverse, and multicultural. The view of the world and its systems as stable and predictable has been replaced with uncertainty, nonlinearity, and chaos. We are experiencing an increasing interdependency among individuals, organizations, communities, nations, and the biosphere.

Succeeding under these conditions will call for changes in how organizations function and, by extension, how leaders lead.

Dynamic, organic, and fluid. As with other dynamic and organic systems, organizations must learn to quickly and smoothly realign structures, processes, and relationships in response to a dynamic external environment.[1]

Chaotic, emergent, and complex. Learning to manage chaos and complexity, the enemies of the old order, will set twenty-first century leaders apart from their predecessors. Business can draw on chaos theory and complexity sciences to gain insights into the nature of organizations as complex adaptive systems.[2]

Holistic, quantum, and integral. Quantum mechanics' revelations about the unpredictable and seemingly random behavior of subatomic particles have led to such technological advances as computers, the Internet, and laser surgery. Recent research in psychology, biology, and neurophysiology suggests that human beings are, indeed, quantum beings. The holistic and quantum implications for leadership are that tapping a person's full potential requires acknowledging, engaging, and integrating the whole person—body, mind, and spirit—including bridging the analytical and artistic sides of workers to increase their creativity.

Adaptive, flexible, and agile. The dynamic and chaotic nature of the business and social environment, the relational and networked nature—and inherent interdependence—of the knowledge economy, the need to quickly recognize and move on new opportunities in a dynamic environment—all these conditions call for adaptive and flexible leadership styles, organizational and personal agility, and far more collaboration than in the past.

Visionary, insightful, and futuristic. Creativity, insight, vision, and integrative capacity—the ability to link ideas together—will be important for bridging the economic, digital, and social divides, as well as other resource gaps, and for solving ecological problems and other pressing global issues. Businesses that possess these capabilities will be able to see these problems and imbalances as new opportunities for responsible and sustainable business.

Leading an organization to succeed in this new context is a paradigm shift for leadership as shown in Exhibit 1, a clear movement away from the old model that has informed leadership behavior and development for most of the twentieth century.

Exhibit 1. The Changing Paradigms of Leadership

Old Paradigm		New Paradigm
■ Self-Centered		**■ Community-Centered**
Ethnocentric	⇨	Community-oriented
Individualistic	⇨	Communitarian
Authoritative	⇨	Collaborative/shared
Short-term interest	⇨	Service to community
■ Old Sciences		**■ New Sciences**
Newtonian	⇨	Quantum
Linear	⇨	Nonlinear
One truth	⇨	Multiple truths
Reductive	⇨	Emergent
■ Uniformity		**■ Diversity**
Hierarchical	⇨	Lateral
Absolute perspective	⇨	Contextualism
Selective	⇨	Inclusive
Simplicity	⇨	Complexity
■ Pure Rationality		**■ Positive Intuition**
Actuality	⇨	Potentiality
Intellectual stimulation	⇨	Emotional arousal
Problems	⇨	Opportunities
Conservative	⇨	Creative
■ Partial		**■ Impartial**
Atomistic	⇨	Holistic
Exclusionary	⇨	Synergistic
Analysis	⇨	Synthesis
Partial	⇨	Integral

Old Paradigm		New Paradigm
■ Profit Orientation		**■ Multiple Orientations**
Theory X	⇨	Theory Z
Competition	⇨	Cooperation
Economic	⇨	Social, environmental, economic
Profit-oriented	⇨	Triple bottom line
■ Certainty		**■ Uncertainty**
Clarity	⇨	Ambiguity
Order	⇨	Chaos
Determinate	⇨	Indeterminate
Stability	⇨	Change
■ Command & Control		**■ Flexibility & Empowerment**
Top down	⇨	Egalitarian
Controlling	⇨	Inspiring
Doubtful	⇨	Trusting
Domination	⇨	Collaboration
■ Old Metaphors		**■ New Metaphors**
Clockwork/machine	⇨	Brain/living ecosystem
Static box	⇨	Dynamic flow
Solid ice	⇨	Emergent cloud
Building	⇨	Web/network

THE NEW ROLES OF A LEADER

As they face more complexities, competition, and change than at any other time in history, organizations that thrive in the new century—actually, in the new world—will be those that learn capabilities and disciplines especially suited to coping with the new conditions. Sources of competitive advantage will derive from an organization's ability to practice ethics and social responsibility, collaboration, relationship building, creativity and innovation, adaptation and flexibility, and systems thinking, as well as to manage complexity and thrive in chaos.

To effectively guide their organizations in learning and applying these new capabilities and disciplines, successful leaders will perform three new creative roles or functions implicit in the new leadership paradigm shown in Exhibit 1:

- Social artist
- Spiritual visionary
- Cultural innovator

Social artists. Social artists are individuals who continually work on themselves to develop skills to provide consultation, leadership, and guidance on changing paradigms, values, laws, and structures of their societies and organizations. Social artists work in collaborative networks to create social innovation. They help cultures and organizations to move from patriarchy to lateral partnership, from dominance to circular sharing. Social artists help people to envision, discover, and realize the most beautiful, powerful, and evolutionary of the possibilities—one that evokes a better world that works for everyone.

Spiritual visionaries. Spiritual visionaries are individuals who interpret the universe and people's roles therein—or in the case of an organization, its role in business and society and its employees' roles in fulfilling its mission. They articulate with authority, eloquence, and depth of insight. They provide deeper meanings, inspiration, and fresh insights about the human condition. They create and utilize powerful visions, metaphors, and symbols. They are the gateways for humanity to explore new facets of the future, to explore collective consciousness. They pioneer new, dynamic, and flexible ways of thinking about holistic problems and questions of the world. They embody and model the search for wholeness, unity, completeness, love, peace, and fulfillment.

Cultural innovators. Cultural innovators are individuals who are advocates and pioneers of new ideas, values, artifacts, and lifestyles in society or the world of work. They explore and nurture creative talents and abilities of people to create change. They are radicals and trendsetters who bring ideas home by making them palatable to their people. They revive a sense of civic vibrancy or engagement lacking in the lives of many people whose societies and organizations are in transition and who find themselves negotiating between tradition and modernity without the intellectual resources to make sense of it all.

Some exceptional leaders of the past may have performed one or more of these functions, to one degree or another. However, the changing times require that these now be the expectations for competent leaders, part of the job description.

A NEW MIND-SET AND HEART-SET

These three creative roles for leaders embody the paradigm shift in leadership values and practices shown in Exhibit 1. They imply nothing less than a dramatic change in the leadership mind-set; recognition of the heart as the seat of critical leadership capabilities; and the integration of mind and heart to develop a "holistic toolkit"—a compilation of both mind-set and heart-set.

Mind-set refers to the rational and intuitive powers—

effective thinking patterns, intellectual skills, perceptions, insights, perspectives, and attitudes—that enable a leader to recognize, conceptualize, and deal with the scope, nature, and complexities of the new era. Heart-set refers to emotional powers—values, determination, motivation, will, passion, belief, grit— that enable a leader to desire and then tenaciously pursue, sometimes in the face of considerable obstacles, what he or she wants or believes to be right.

The following are eleven domains that define the heart-set and mind-set of leaders for the twenty-first century. Each of them is defined through examples that are illustrative but certainly not exhaustive.

- *Creativity and intuition:*

— Develops and uses creative and intuitive abilities; sees intentionally, knows intuitively.

— Emphasizes creativity and transformation; uses creative tension to foster change and

new ideas; finds new ways to bridge complex problems and gaps.

— Recognizes the need for real change that matters to people's enduring needs.

- *Passion and inspiration:*

— Feels vitally alive, is passionate, lives with great enthusiasm; has imagination and hope for the future; trusts life's process.

— Brings heart, soul, and spirit to work.

— Makes a difference in the world by reaching out, touching, and hopefully even inspiring others; enables others; evokes the possible, offers the lure of "becoming," and shows others their own giftedness.

- *Meaning and reflection:*

— Engages the heart, discovers deeper meaning, serves a higher purpose; acts from purpose and meaning; engages in continuous self-development and reflection, utilizing reflective, artistic, and relational spaces and reflecting through music and art.

— Draws on qualities of empathy, spirit, pattern recognition, and a rich emotional complexity to create meaning for others; calls others to service in ways that link inner and outer realities, universal plans, and passionate commitment.

— Develops meaning and insight through individual and collective reflection; cultivates ritual and celebration; acknowledges and uses mistakes as opportunities to learn, reflect, and forgive.

- *Vision and insight:*

— Has a clear vision of the future he/she wishes to create; uses a long-term perspec tive, thereby creating viability for current and future generations; sees the trends and the emergence of new patterns; perceives linkages that will generate new opportunities and generative insights; sees horizons rather than borders.

— Holds a shared vision and core values in trust with others, working together to operationalize them; includes diverse individuals and perspectives to develop insight.

— Is a paradigm pioneer, helping people and organizations find their vision.

— Facilitates self-organizing, self-regulating, and self-renewing systems.

- *Courage and accountability:*

— Takes risks, tackles the difficult issues, initiates change.

— Willing to assume full responsibility for one's decisions and actions; unwavering courage; self-control.

— Keeps a sense of justice (and fairness) towards others; challenges others when they depart from core values held in trust.

— Transforms self, groups, and institutions.

- *Integrity and authenticity:*

— Practices deep listening; develops principled and thoughtful thinking.

— Walks the talk; stands up for what is right with ultimate integrity.

— Is honest with self and others; is consistent and sincere; shows mutual respect.

- *Openness and flexibility:*

— Creates and maintains a free flow of information; fosters and demonstrates open-mindedness.

— Facilitates fluidity and flexibility in group processes and structures; creates collaborative networks; moves easily between cultures (within organizations).

— Becomes a steward of the process of change.

- *Wisdom and consciousness:*

— Develops an integrated mind and self-awareness, including awareness of own ethics and values; nurtures soulfulness.

— Sees the big picture—the shift in consciousness and the cultural, political, economic, and social transformation within individuals and collectives.

— Conscious of interconnectedness and holism; sees the long-term implications of decisions and actions on the system; emphasizes social responsibility toward the community, society, and environment.

— Promotes and facilitates reflection by others, creating safe spaces for them to step back and think about the meanings and consequences of what they are doing.

- Stewardship and care:

— Develops empathy and understanding towards other people by taking their feelings into consideration; develops trusting relationships, listens with attention; emphasizes empowerment, delegation, and cooperation; seeks to be in service rather than in control.

— Attends to the well-being (basic needs and human rights) of others and provides opportunities for them to sustain themselves.

— Creates communities of reciprocal care and shared responsibility where every person matters and each person's welfare is the common concern; creates a supportive environment where people can thrive, grow, and live in peace with one another, with strong trusting relationships forged by visioning, leading, learning, and acting together.

- *Growth and development:*

— Strengthens and sustains individual growth and self-actualization.

— Deeply commits to the personal, professional, and spiritual growth of each individual in the institution/organization; discovers, coaches, and nurtures each person's untapped potential; expects the best from people.

— Creates opportunities for people to experience success (efficacy); celebrates individual and group success.

— Promotes group and community capacity building and progress; creates lifelong teaching and learning communities.

- *Harmony and balance:*

— Achieves balance in the emotional, spiritual, and physical aspects of life.

— Understands the interdependent relationship between human and natural systems, and works to enhance their viability; promotes harmony with nature, and provides sustainability for future generations.

— Values diversity and inclusiveness; fosters diversity by respecting different perspectives; establishes and sustains inclusiveness of stakeholders.

— Generates and supports interdependent and interdisciplinary group processes.

— Builds bridges rather than barriers among peoples, encouraging and practicing dialogue; generates and sustains peace among others; aids peace efforts globally; recognizes and promotes the spiritual connectedness of all life.

— Views most situations from a more integrated position; develops a holistic systems perspective, taking the whole system into account with all its cyclicality, interdependence, and complexity; generates and supports holistic thinking as a basis for action.

This, then, is the art, craft, and science of twenty-first century leadership—world-making, spirit-catching, mind-growing, soul-quaking leadership. Leaders who possess such intellectual and emotional capacities, resources, and skills can awaken the organizational spirit. They are catalysts for individual growth, helping each employee tap into the boundless human potential for personal, organizational, and global transformation. Simultaneously they can engage a wide range of stakeholders in visioning and implementing innovative and integral solutions for the world's most imperative and complex problems.

THE IMPLICATIONS FOR TODAY'S LEADERS

The challenges our organizations, our world, now face would strongly suggest that what we believed in the past to be our strengths as managers are inadequate—and in the worst case, liabilities—for moving forward. As managers we have been too preoccupied with short-term profit, material success, speed, efficiency, divisions, and specialization. We parsed management knowledge into managerial functions, disciplines, and further subdisciplines. We built our organizations around formal hierarchies and linear structures. We drew on predictive, cause-and-effect models of human and organizational behavior. In focusing too heavily on analysis, we ignored synthesis. In using models that are too rigid, isolated, specialized, formal, and unconnected, we created structures and processes that stifle the needs of our team members for meaning, reflection, exploration, creativity, risk taking, and connection. In emphasizing problems and problem solving, we neglected to nurture hope, discovery, and imagination. In making numbers, statistics, and material performance the means and the goal, we neglected our collective need for soulful engagement, inspiration, qualitative inquiry, pattern formation, narrative, and meaning. We fragmented our values, our behavior, our families, our spirituality, our lives, and our work. We separated our bodies from our minds, our minds from our hearts, our hearts from our soul, and our souls from one another.

We suffer now from the lack of individuals equipped to perform the new functions we need from our leaders. This is not only a matter of the wrong leader mind-set for taking in and responding to the realities of global change but also, and even more tragically, a lack of the positive values, integrity, self-awareness, and deeper consciousness that nourish a leader heart-set.

Business leadership in the twenty-first century will be about embracing employees—in actuality, all stakeholders—as whole persons, acknowledging not only their cognitive faculties but also their social, emotional, physical, and spiritual faculties, to engage their hearts and spirits as well as their minds. To do this, leaders must be dedicated and passionate about making a positive difference in the lives of people, which can only grow from authentic enthusiasm, love, and concern.

For many of us, the road to becoming this kind of leader will involve some degree of personal growth—even transformation. Only by learning to tap our own undeveloped capacities and potential as human beings can we show our employees and organizations how it is done.

As leaders we also have the responsibility to create supportive learning environments built on connection, coherence, mutually created meaning, dynamic relationships, and an integrated systems perspective. In these kinds of environments employees can develop the values, perspectives, and capabilities that will enable them to work in the new ways. In these kinds of environments we will grow this century's leaders.

Notes

1. The basic concept of dynamic and organic systems is that all things tend to organize themselves into patterns, e.g., ant colonies, immune systems, and human cultures; furthermore, they go through cycles of growth, mass extinction, regeneration, and evolution.

2. Chaos theory encompasses a set of ideas that attempt to reveal structure in aperiodic, unpredictable dynamic systems such as cloud formation or the fluctuation of biological populations. Nonlinear dynamic systems are those where the relationships between time dependent variables are nonlinear. For example, according to the butterfly effect, small differences in initial conditions can quickly lead to large differences in the future state of a system. Chaos theory and complexity sciences provide insights into the nature of complex adaptive systems, those that respond to both feedback and feed forward and are operating in turbulent environments.

ACKNOWLEDGMENT

The author would like to thank Professor Nancy Adler for the inspiration she provided, without which this article would not have reached its potential.

ABOUT THE AUTHOR

Fahri Karakas is a researcher on the Faculty of Management at McGill University, in Montreal, Quebec, Canada, who specializes in organizational behavior. His research interests include values and spirituality in the workplace, personality, leadership, group dynamics, educational leadership, social innovation, complexity, chaos theory, positive organizational scholarship, and appreciative inquiry.

From: Fahri Karakas, "The Twenty-First Century Leader: Social Artist, Spiritual Visionary, and Cultural Innovator," *Global Business & Organizational Excellence,* (March/April 2007): 44-50.

16

CHAPTER 16

STRATEGIC COMMUNICATIONS & NEGOTIATION

Communication goes beyond knowing about sentence structure, the parts of a staff briefing, and transition phrases to use in a speech. While chapters 2, 3, and 8 gave you an overview of specific communication techniques, this chapter will introduce you to the use of communication for negotiation and diplomacy in the strategic environment. As you studied in chapter 6, conflict between humans and in organizations is inevitable. The bridges that resolve conflicts are negotiation and diplomacy.

INTRODUCTION

The first article, "Principles of Strategic Communication," will introduce you to basic terminology and characteristics of strategic communications. It was developed by the Department of Defense to assist in their quest to develop policy and doctrine for strategic communication concepts.

Whether you're presenting a speech at school, a new training course for your squadron, or a news release for a CAP activity, your goal is to make your idea so memorable that your audience will act on or respond to the information they've received. The authors of the second article, "Ideas that Stick," illustrate six principles of communication that help ensure that your audience will remember your message.

Even if you have a great idea and follow the steps to make your idea sticky, automatic acceptance is not guaranteed. From reaching a final decision at a CAC meeting to persuading your encampment staff to adopt a new plan of action to setting your work schedule with your boss, negotiation skills are necessary. In the next article, "The Art of Negotiation," the author shares practical tips for effective negotiation.

In the fourth article, "Negotiating Effectively Across Cultures: Bringing Out the DEAD," the author presents a framework for understanding and preparing for negotiating with individuals from other cultures. The skills presented in the article are not restricted to cross-cultural communication; you may also find this framework useful for communicating with peers from your own culture.

When negotiation is not successful, disputes can follow. This highlights the need for diplomacy. Disputes don't have to be as complex as the search for peace in the Middle East for diplomatic techniques to be useful. In "Preventive Diplomacy," you will read about a middle school curriculum developed by the author to instruct students in conflict management and prevention. The skills of negotiation and principles of preventive diplomacy covered in this article can be useful for cadet officers across the range of their daily interactions, from personal relationships to unpopular command decisions.

The final article takes several steps back from a focus on the cadet squadron and personal skill development, addressing the use of negotiation, creative ideas, and diplomacy at the international level. In "The Not-So-Black Art of Public Diplomacy," the author describes the importance and challenges of public diplomacy in US foreign relations. These concepts will be important to cadets who are considering careers in public service, international relations, and military service.

CHAPTER OUTLINE

This chapter's readings are:

Principles of Strategic Communication
Department of Defense, "Principles of Strategic Communication," (August 2008).

The Art of Negotiation
Brenda Goodman, "The Art of Negotiation," *Psychology Today* 40, no. 1 (January/February 2007): 64-65.

Negotiating Effectively Across Cultures
John W. Miller, "Negotiating Effectively Across Cultures: Bringing Out the DEAD," Squadron Officer College SI-5120 Student Reading (February 2010).

Preventive Diplomacy: Training a New Generation for Peace
Carl Hobert, "Preventive Diplomacy: Training a New Generation for Peace," *Independent School Magazine* (Spring 2008).

The Not-So-Black Art of Public Diplomacy
Humphrey Taylor, "The Not-So-Black Art of Public Diplomacy," *World Policy Journal* 24, no. 4 (Winter 2007/08): 51-59.

CHAPTER GOALS

1. **Summarize key principles of strategic communications.**
2. **Appreciate the value of diplomacy in preventing and resolving conflict.**
3. **Describe principles of negotiation.**

16.1 Principles of Strategic Communication

By the Department of Defense, Office of the PDAS for Public Affairs

OBJECTIVES:

1. Define the term "strategic communication" and explain its relevance to leaders.
2. List and describe the nine principles of strategic communication.

Principle: *A fundamental tenet; a determining characteristic; an essential quality; an enduring attribute.*

Strategic Communication (SC) has been described as the orchestration and/or synchronization of actions, images, and words to achieve a desired effect, yet there is more to understanding the concept.

As the joint forces and agencies of the U.S. Government have begun executing Strategic Communication processes, common fundamentals have emerged. Through the collaborative efforts of DoD, State Department, civilian educators, and Strategic Communication practitioners, those common fundamentals have been consolidated and refined into nine principles of SC, described below. These principles are provided to assist dialogue and instruction promoting understanding of Strategic Communication.

Shown below are nine principles of SC, with a short description of each. A more detailed explanation of each principle follows. The principles are not listed in any order of precedence.

Leadership-Driven
Leaders must lead communication process

Credible
Perception of truthfulness and respect

Dialogue
Multi-faceted exchange of ideas

Unity of Effort
Integrated and coordinated

Responsive
Right audience, message, time, and place

Understanding
Deep comprehension of others

Pervasive
Every action sends a message

Results-Based
Tied to desired end state

Continuous
Analysis, Planning, Execution, Assessment

LEADERSHIP-DRIVEN. Leaders must decisively engage and drive the Strategic Communication process. To ensure integration of communication efforts, leaders should place communication at the core of everything they do. Successful Strategic Communication – integrating actions, words, and images – begins with clear leadership intent and guidance. Desired objectives and outcomes are then closely tied to major lines of operation outlined in the organization, command or join campaign plan. The results are actions and words linked to the plan. Leaders also need to properly resource strategic communication at a priority comparable to other important areas such as logistics and intelligence.

CREDIBLE. Perception of truthfulness and respect between all parties. Credibility and consistency are the foundation of effective communication; they build and rely on perceptions of accuracy, truthfulness, and respect. Actions, images, and words must be integrated and coordinated internally and externally with no perceived inconsistencies between words and deeds or between policy and deeds. Strategic Communication also requires a professional force of properly trained, educated, and attentive communicators. Credibility also often entails communicating through others who may be viewed as more credible.

UNDERSTANDING. Deep comprehension of attitudes, cultures, identities, behavior, history, perspectives and social systems. What we say, do, or show may not be what others hear or see. An individual's experience, culture, and knowledge provide the context shaping their perceptions and therefore their judgment of actions. We must understand that concepts of moral values are not absolute, but are relative to the individual's societal and cultural narrative. Audiences determine meaning by interpretation of our communication with them; thus what we say, do, or show may not be what they hear or see. Acting without understanding our audiences can lead to critical misunderstandings with serious consequences.

Understanding subjective impacts of culture, language, history, religion, environment, and other factors is critical when crafting communication strategy for a relevant population. Building relationship and collaboration with the interagency, coalition, host nation, academic, non-profit, and business communities can facilitate better understanding of audiences.

DIALOGUE. Multi-faceted exchange of ideas to promote understanding and build relationships. Effective communication requires a multi-faceted dialogue among parties. It involves active listening, engagement, and the pursuit of mutual understanding, which leads to trust. Success depends upon building and leveraging relationships. Leaders should take advantage of these relationships to place U.S. policies and actions in context prior to operations or events. Successful development and implementation of communication strategy will seldom happen overnight; relationships take time to develop and require listening, respect for culture, and trust-building.

PERVASIVE. Every action, image, and word sends a message. Communication no longer has boundaries, in time or space. All players are communicators, wittingly or not. Everything the Joint Force says, does, or fails to do and say has intended and unintended consequences. Every action, word, and image sends a message, and every team member is a messenger, from the 18-year-old rifleman to the commander. All communication can have strategic impact, and unintended audiences are unavoidable in the global information environment; therefore, leaders must think about possible "*N*th" order communication results of their actions.

UNITY OF EFFORT. Integrated and coordinated, vertically and horizontally. Strategic Communication is a consistent, collaborative process that must be integrated vertically from strategic through tactical levels, and horizontally across stakeholders. Leaders coordinate and synchronize capabilities and instruments of power within their area of responsibility, areas of influence, and areas of interest to achieve desired outcomes. Recognizing that your agency/organization will not act alone, ideally, all those who may have an impact should be part of communication integration.

RESULTS-BASED. Actions to achieve specific outcomes in pursuit of a well-articulated end state. Strategic communication should be focused on achieving specific desired results in pursuit of a clearly defined end state. Communication processes, themes, targets and engagement modes are derived from policy, strategic vision, campaign planning and operations design. Strategic communication is not simply "another tool in the leader's toolbox," but must guide all an organization does and says; encompassing and harmonized with other functions for desired results.

RESPONSIVE. Right audience, right message, right time, and right place. Strategic Communication should focus on long-term end states or desired outcomes. Rapid and timely response to evolving conditions and crises is important as these may have strategic effects. Communication strategy must reach intended audiences through a customized message that is relevant to those audiences. Strategic Communication involves the broader discussion of aligning actions, images, and words to support policy, overarching strategic objectives and the longer term big pictures. Acting within adversaries' decision cycles is also key because tempo and adaptability count. Frequently there will be a limited window of opportunity for specific messages to achieve a desired result.

An organization must remain flexible enough to address specific issues with specific audiences, often at specific moments in time, by communicating to achieve the greatest effect. All communication carries inherent risk and requires a level of risk acceptance within the organization. Leaders must develop and instill a culture that rewards initiative while not overreacting to setbacks and miscues. While risk must be addressed in the form of assumptions in planning, it should not restrain leaders' freedom of action providing it has been taken into considerations appropriately.

CONTINUOUS. Diligent ongoing research, analysis, planning, executing, and assessment that feeds planning and action. Strategic Communication is a continuous process of research and analysis, planning, execution, and assessment. Success in this process requires diligent and continual analysis and assessment feeding back into planning and action. Strategic Communication supports the organization's objectives by adapting as needed and as plans change. The SC process should ideally operate at a faster tempo or rhythm than our adversaries.

From: Department of Defense, Office of the Principal Deputy Assistant Secretary for Public Affairs, August 2008.

16.2 The Art of Negotiation

By Brenda Goodman

OBJECTIVES:

3. Define the term "negotiation."
4. Recall the five core concerns in creating disputes and finding resolution.
5. Name the most important element of effective negotiation.
6. List several ways to improve negotiation.

Have you ever had a discussion with yourself about when to go to bed? The word "negotiation" may conjure thoughts of hostage standoffs and high-stakes labor disputes, but there's a more quotidian brand of conflict resolution that enters daily life at nearly every turn. Negotiation, in fact, doesn't necessarily even require another person.

Mary P. Rowe, an ombudsman at MIT, encourages people to think of negotiation as "all interactions between two or more points of view; it's possible to negotiate with yourself."

Negotiations crop up on the way to decisions big and small—when to fill the gas tank, how to spend money, who picks up the kids, whether to get married.

Granted, forging a compromise over which DVD to watch isn't the same as signing the Camp David Accords, but regular human beings can benefit from the same skills world leaders use to solve problems. And best of all, getting better at reaching agreement is pretty painless.

Principled negotiation is a strategy that seeks to move both parties away from polarizing and usually entrenched positions, and into the realm of interests. It asks how both parties can get their interests satisfied while keeping their relationship strong. Negotiating well means neither party need feel cheated, manipulated, or taken advantage of.

Psychologist Daniel L. Shapiro, associate director of the Harvard Negotiation Project, has trained Palestinian and Israeli negotiators. He taught members of the Serbian parliament how to negotiate. Unfortunately, he reports, none of that has given him any additional clout at home.

When he was dating his wife, Mia, a painful imbroglio erupted after he asked her to watch his apartment while he was away. He returned to discover she had redecorated. Gone was his "cool" construction lantern. The card table he ate on had a new flowered tablecloth.

"In truth, it looked better," but Shapiro was incensed. The trouble, he recognized later, was that Mia had inadvertently trampled his autonomy. That turns out to be one of five "core concerns" his research identifies as critical in creating disputes and finding resolution. He defines autonomy as a person's freedom to make decisions for himself.

The other core concerns are appreciation, or having actions acknowledged; affiliation, being treated as a colleague; status, feeling that others respect one's standing; and having roles and activities that are fulfilling. Cross one of the needs and conflict arises. Respect them, and compromise is around the corner.

The most important element of effective negotiation, says Rowe, is preparation, preparation, preparation. She recommends drafting a letter that includes an objective statement of the facts, explains how those facts were injurious, and outlines what the writer thinks should happen next. Even if the letter is never sent, writing it can help clarify what is needed to repair any damage.

If there is not enough time for a letter, even a 10-minute break from a highly charged situation allows murky issues to be thought through and real needs to come to light. Advises Shapiro: "Take those core concerns and write them on a piece of paper. Figure out which of them are being violated for you and for the other person."

KEY PRINCIPLES

- **Listen First** "There's a saying among negotiators that whoever talks the most during a negotiation loses," says Bobby Covic, author of Everything's Negotiable! Being the first one to listen is crucial to building trust. Just getting the listening part of a negotiation right can satisfy many of the core concerns Shapiro cites.

However, listening—really paying attention to what the other person has to say—is hard. Gregorio Billikopf, a negotiator for the University of California system, offers several good listening practices:

- **Sit Down** This signals to the other person that time will be spent to hear their side. Never ask someone to talk if there isn't enough time to listen.
- **Find Common Ground** Approach the other person by

talking about a neutral topic of mutual interest—say, baseball or knitting. It helps both parties relax and starts the flow of conversation. Transition to the problem by saying, "I want to talk about an issue important to me, but first I want to hear what you have to say about it."

- **Move In** Leaning in to the conversation indicates interest. Head nods also help in letting the other side know their thoughts are being followed. But constant nodding or saying "right" over and over will seem insincere.

- **Keep Your Cool** Experts agree on ground rules for communicating problems— no yelling and no walking away.

- **Be Brief** Don't go on and on, says Billikopf. He also suggests avoiding words such as "we disagree," a phrase that throws a person to the defensive.

- **Forget Neutrality** Trying to control your emotions usually backfires, says Shapiro. The other person can read anger and frustration in a wrinkled forehead or a tense mouth, and negative emotions ruin negotiations. Instead, mine the situation to find whatever positive emotions can be brought to the table—like letting a spouse who's fallen behind on his end of the chores know that his hard work is admirable and the extra money he's earning is appreciated.

- **Avoid Empty Threats** Intimidation can be powerful—but use it sparingly. Empty threats will diminish the other person's respect for you.

- **Don't Yield** Caving on important issues may seem noble, says Billikopf, but it ruins a relationship. "You're not asking the other person to consider your point of view," he says. Instead, look for compromises. Compromise is like stretching. Stop doing it and pretty soon there's no way to bend at all.

THE GENDER GAP

Ask a man to describe negotiation and he's likely to compare it to a ball game or a wrestling match. Women, on the other hand, find it more like going to the dentist.

By a factor of 2.5, more women than men feel a "great deal of apprehension" about negotiating, reports economist Linda Babcock, of Carnegie Mellon. Women go to great lengths to avoid the bargaining process—paying almost $1,400 more to avoid negotiating the price of a car. (That may explain why 63 percent of those who buy cars made by Saturn, a company that promises a no-haggle price, are women.) But "failing to negotiate her salary just once will cost a woman $500,000 over the course of her career," she says.

Babcock suggests three things for women to get more of what they want:

- **COMMIT** "Given that 20 percent of adult women say they never negotiate at all, the most important thing to do is to decide to use negotiation in the first place," she says.

- **PRACTICE** Negotiate little things, even crazy items that are never bargained for, like the price of fish at the fish market. As with most behaviors, she says, it gets easier the more you do it.

- **GET TO 'NO'** If you never hear "no" when you negotiate, you haven't asked for enough.

ABOUT THE AUTHOR

Brenda Goodman is an Atlanta-based freelance writer.

From: Brenda Goodman, "The Art of Negotiation," *Psychology Today* 40, no. 1 (January/February 2007): 64-65. Used with permission.

16.3 Negotiating Effectively Across Cultures: Bringing Out the DEAD

By John W. Miller, Ph.D.

OBJECTIVES:

7. Define the term "negotiation."
8. Contrast distributive deals with integrative deals.
9. Define the term "culture."
10. Define the term "institution."
11. Define the term "reframing."
12. Define the concept of "thin-slicing."
13. List and briefly define four general cultural patterns.
14. List and briefly define four conflict styles.

According to Jeanne Brett in her book *Negotiating Globally*, a negotiation is a communicative "process through which people with conflicting interests determine how they are going to allocate resources or work together in the future."[1] Negotiations can range from a mundane discussion about where to eat lunch to an intricate arms treaty with implications for all of humanity. Most of our negotiating experiences are of the more commonplace variety. Yet for military leaders, the ability to negotiate effectively is no mean skill as the success or failure of the process can have an impact on large groups of people. You may never have an opportunity to mediate an arms agreement or broker a multi-billion dollar contract, but as a leader, you must constantly use negotiation skills as part of your daily work routine. Examples abound. Someone wants to take annual leave while your work group is in the midst of a high-visibility project. How will you handle the request? Another person comes to you in confidence to explain how he cannot work with a fellow officer on the same key project. What negotiation skills do you use? These interactions present their own difficulties when enacted within the framework of our own cultural and organizational norms. Consider the added challenge when conducting such negotiations across cultures.

Although this article will provide a brief overview of negotiation, its primary purpose is to introduce you to a framework for understanding the intricacies of negotiating cross-culturally. The information introduced here cannot by itself make you a competent negotiator. What it can do is prepare you for your next negotiation by giving you insight into the issues and interests at play in both intra-cultural and cross-cultural negotiations.

WESTERN-STYLE NEGOTIATING: MAKING THE DEAL

When we think of a negotiation, most of us think of what we commonly call "The Deal." Deals, as they are often referred to in the U.S. and other Western nations, are agreements or settlements reached after a discussion over an issue. We have all made deals at one time or another. In North America and Northern European contexts, these negotiations can be divided into two groups, distributive and integrative.

Distributive Deals

A distributive deal is one in which two people negotiate over a single issue and the issue is often the cost of an item. If you have ever haggled over the price of a used car at your local dealership or a knick-knack at a flea market, you have engaged in distributive deal-making. The distributive deal is what most people around the world associate with negotiation. The salesperson or shopkeeper starts high, the customer counters low, and the dance goes on until either an agreement is reached or the customer walks away. The term "distributive" refers to the way in which the resources will be distributed. In the shopkeeper/customer scenario the resources being distributed are money and a product. Each side takes a position and dickers back and forth until a mutually acceptable price or compromise is reached and the goods are transferred. This kind of deal works well when positions and interests are well defined. Yard sales and used car lots are common distributive deal-making situations. Although one can reduce all negotiation to resource distribution, the portrayal is overly simplistic. As Brett emphasizes, "distribution is only one aspect of negotiation."[2]

Integrative Deals

To explain the intricacies of negotiation, experts in the field often tell the parable of the orange. In this time-worn tale, two young sisters are in the kitchen arguing over the last orange setting in a bowl on the kitchen table. The resource in this case is the orange and both sisters want it. This story appears to be another example of a distributive deal, but with a zero-sum outcome. There is only one issue and one resource—the orange. As the argument escalates Mother enters the kitchen, listens for a second, takes the orange, cuts it in half, and distributes an equal share to each sister. Mom has once more displayed the Wisdom of Solomon—or perhaps not. Both sisters are still unhappy and another argument quickly ensues.

How might the girls or their mother have brought a happy ending to this classic tale? The answer is through talk. Real negotiation usually demands more than simply taking a firm position, such as haggling over price or arguing over who has the better claim to a limited resource. Yet in most negotiations our natural inclination tends to be similar to that of the two sisters—take a position and stick to it. In an integrative deal, each sister—or the mother as a third-party intermediary—would have drawn out the interests of the other side. The term "interests" refers to the reasoning behind a position such as the negotiator's needs, fears, or concerns. In this case, each sister's position is based on her individual need. And there lies the hidden rub: Each had a different need. One wanted to bake an orange cake and needed only the rind. The other just wanted to drink a glass of orange juice and cared only about the juice. If the sisters had taken the time to talk to each other and revealed their interests, they could have divided the orange in a way that would have fully satisfied both their needs. Integrative deal-making seeks to expand the resources beyond those available. This type of negotiation requires trust and the willingness to reveal more rather than less to the parties involved. Inevitably, an integrative deal is much more difficult when negotiating across cultures because it focuses on need rather than commodities. These needs are known as "interests."

Focus on Interests

According to Brett, the key to uncovering interests is by "asking *why* and *why not*."[3] In the early 1980s, Roger Fisher and William Ury first introduced the concept of interest-based negotiations to a wide audience. In their seminal work, *Getting to Yes*, they described a well known integrative settlement to illustrate such a deal: The Camp David Accords signed by Egypt and Israel.[4] The initial positions of President Sadat of Egypt and Prime Minister Begin of Israel were completely at odds. Among other demands, Sadat wanted Israel to turn over the entire Sinai Peninsula. Begin refused to return to the same situation that existed before the 1967 war. During the negotiation process, each side would redraw the map and pass it to the other side, and each time the other side would reject it. Like the two sisters, neither side would budge. Fisher and Ury described the situation in this way:

> Israel's interest lay in security; they did not want Egyptian tanks poised on their border... Egypt's interest lay in sovereignty; the Sinai had been part of Egypt since the time of the Pharaohs. After centuries of domination by Greeks, Romans, Turks, French, and British, Egypt had only recently regained full sovereignty and was not about to cede territory to another foreign conqueror.[5]

In his memoir, *Keeping Faith*, President Jimmy Carter described those historic negotiations in great detail.[6] According to him, any attempt to split the Sinai would have resulted in the collapse of the talks. Through Carter's persistent mediation, Sadat and Begin were able to look past their positions and focus on their interests. Return of the Sinai was Sadat's primary need. Having a military presence there was not. On the other hand, Israel's top interest was security, not Sinai real estate. The issue of land was important for Begin, too, but not his first priority. After much discussion, he agreed to remove Israeli settlements from the Sinai, contingent upon Knesset approval.[7] Both sides eventually agreed to a plan that allowed the Egyptian flag to fly above the Sinai, while Egyptian "tanks would be nowhere near Israel."[8]

As Sadat and Begin's interests involving the Sinai illustrate, integrative deals require a negotiation process that requires a clear understanding of one's own interests and the willingness to prioritize them. In a negotiation where many issues are in play, one must expect that not all interests will be met on either side. The process of identifying and prioritizing interests demands careful planning. Yet Don Conlon, Eli Broad Professor of Management at Michigan State University, cites inadequate planning as the biggest mistake made in negotiations.[9] Cross-cultural negotiation requires even greater time and planning. The next section introduces the intricacies of brokering deals across cultures.

NEGOTIATING IN CROSS-CULTURAL CONTEXTS

Cultures are complex sets of learned behaviors, beliefs, values, and assumptions. Objective aspects of culture, such as art, architecture, food, music, dress, and language are observable. Aspects such as values, beliefs, and assumptions are subjective and more difficult to discern. These subjective aspects are hidden not only from the sojourner, but from the native as well. Differences in these hidden areas can act as cultural hooks that hang us up and lead to ambiguity, confusion, and misunderstanding.

Behaviors

Your ability to interpret behaviors when negotiating in divergent cultural contexts is important. You may never learn to make the distinctive "snap" that ends a Liberian handshake or use stainless steel chopsticks as deftly as a Korean. And that's OK. What is important is being open to divergent cultural behaviors and withholding judgment. Of course, no matter how hard you try, you will still make mistakes. Once as a member of an American negotiating team in Japan, I was asked to present a proposal to a Japanese university's chief administrators. I felt proficient enough to outline the proposal in Japanese. To lighten the atmosphere, I decided to begin by telling a rather bland joke about jetlag. I practiced until I had it down pat. Much to my chagrin, however, the punch-line was met with stony silence. What I learned afterwards was that in Japan jokes are inappropriate in formal contexts. I would have been better served had I begun with a humble apology for the inadequacies of our proposal. Humility, not humor, is the acceptable opening for such proceedings. Fortunately, in spite of my clumsy introduction, the proposal was accepted. These kinds of cultural mistakes are part of the learning process. Most people will understand if you acknowledge you have made a mistake and seek to make amends.

In fact, such mistakes will occur frequently in any situation where people from diverse cultural contexts collaborate and work together. Coalition teams, for example, provide fertile ground for misunderstandings and conflict. The ability to resolve cultural conflict issues requires patience and openness to differences in behaviors and institutional practices.

Cultural Values, Beliefs, and Assumptions

Cultural values, beliefs, and assumptions are powerful forces within a culture. They are passed down from generation to generation through the family, schools, the media, and religious institutions. Although hidden from our view below the "waterline," these shared concepts are the foundation for all those aspects that are easily perceived. Although it may be convenient to categorize cultures by their values and norms, some caution is needed. In any culture, not all members display or "buy into" these psychological structures. Everyone in a culture is not the same. Therefore, when talking about values, beliefs, and assumptions it is wise to frame them as *generalizations*. It is better to say that Iraqis, for example, "tend to be" collectivist or that American institutions "in general" support individualist values. To do otherwise is to fall into the trap of stereotyping.[10]

Institutions

Institutions, according to Brett, are "economic, social, political, legal, religious institutional environments that effect negotiation."[11] This includes governmental organizations such as the military. Like behaviors, institutional structures are linked to cultural values and beliefs. For example, the fictional nation of Leonia is an Arab Muslim culture situated in the Maghreb region of Northern Africa. Cultures in this region tend to be much more hierarchical than those in the West. Yet there are benefits to this type of organizational structure. In some cases, such hierarchies allow even low level functionaries to have direct access to management at a much higher level than would be common or even acceptable in the U.S. Sometimes it simply means finding out whom to contact to gain access to decision-makers. This requires the forming of alliances, locating third-party intermediaries, and the development of friendships and strong working relationships with host country nationals.

Reframing

When we call someone lazy, we are making a judgment about that person's character. When applied to a group, the judgment has been transformed into a negative stereotype because the attribution is not to just one person, but an entire group and by extension an entire culture. This kind of stereotyping is inappropriate. Before making such sweeping judgments, [you] must clearly define the negative behavior and then determine the cause. The roots of the behavior are more likely tied to values related to cultural domains such as kinship, education, or institutional processes. "Reframing" is a helpful process for moving beyond stereotyping and judgmental language. Stella Ting-Toomey and Leeva Chung, two recognized experts in the field of intercultural conflict resolution, described "reframing" as a communication skill that uses "neutrally toned (to positively-toned) language...to reduce tension and increase understanding."[12] The AFINT instructors could begin the process of understanding the problem by framing their descriptions of behavior in non-judgmental terms:

> ***Judgmental Statement***
> "The students are lazy."
> "The students are unmotivated."
>
> ***Reframed Statement***
> Some students turn in their homework either late or incomplete.
> Some students come to class 5 to 15 minutes late.
> Some students have missed up to three days of class.

Values-Based Negotiation

The story of the sisters and the orange highlights the importance of understanding the interests of all parties concerned. And, as previously stated, interests are the underlying reason for entering a negotiation. Brett's suggestion for discovering the interests of the other party is to ask the questions "Why?" and "Why not?" But Brett also cautioned that "such direct questioning might not work everywhere in the world."[13] In many other cultures, asking direct questions is seen as aggressive and intrusive behavior.

When engaging in cross-cultural negotiation, one is better served by uncovering both values and interests. Quite often the two are entwined. As John Forester pointed out in his article "Dealing with Deep Value Differences," "values run deeper than interests."[14] He goes on to explain how interests—such as time or money—are shed more easily than cultural values because:

When we give up something we value, we often feel we give up part of ourselves, and that's very difficult, very threatening, and hardly compensated by some gain somewhere else.[15]

If we return to the situation at the Camp David Accords, we can see how closely cultural factors are enmeshed with interests. Carter wrote that "there was no compatibility at all" between Sadat and Begin.[16] Yet with the U.S. president acting as a bridge, Sadat and Begin were able to overcome cultural and political differences. The cultural factors ran deep on both sides. Begin's decision to remove the Israeli settlements was a difficult one for a man whose people had forged a new nation in what they believed to be their Promised Land after centuries of persecution. Carter called this concession "a remarkable demonstration of courage, political courage, on the part of Prime Minister Begin."[17]

As Carter did with Sadat and Begin, [you] would do well to discover the values influencing the institutional and personal behavior causing conflict before [you] commence any formal attempts to resolve the issues.

To summarize, cross-cultural negotiations and conflict resolution require attention to values, beliefs, and other psychological aspects of culture that go hand in hand with a group's specific interests. An understanding of these areas is the key to a satisfactory resolution or agreement.

INTERCULTURAL CONFLICT STYLES

In their book, *Managing Intercultural Conflict Effectively*, Stella Ting-Toomey and John Oetzel, define cross-cultural conflict as:

> The experience of emotional frustration in conjunction with perceived incompatibility of values, norms,...goals, scarce resources, processes, and/or outcomes between...parties from different cultural communities.[18]

Clearly, negotiation and conflict are closely linked. Understanding how conflict is displayed in divergent cultural contexts can benefit planners engaged in cross-cultural negotiations. It also serves as a helpful guide in preparing for any negotiation. This section will introduce you to the phenomena of thin-slicing, mind-blindness, and the ICS-DEAD model of intercultural conflict styles.

Thin-Slicing and Mind-Blindness

In his bestselling book *Blink*, Malcolm Gladwell described the phenomenon of rapid cognition known as *thin-slicing*.[19] Thin-slicing is the human ability to use "our unconscious to find patterns in situations and behavior based on very narrow slices of experience."[20] Thin-slicing is used constantly in human interaction as we read the meaning of a glance or a tone of voice. We also thin-slice our way through disagreements and conflict situations. Although the ability to thin-slice is innate, the patterns that frame our ability to slice and dice are learned.

The inability to thin-slice is a condition common to those suffering from autism. People with autism, according to Gladwell,

> "find it difficult, if not impossible to...[interpret] nonverbal cues, such as gestures and facial expressions or putting themselves inside someone else's head or drawing understanding from anything other than the literal meaning of words."[21]

This is exactly what happens when human beings cross into new cultural terrain. In a cross-cultural situation, this temporarily autistic condition, a mental state that Gladwell calls "mind-blindness," causes us to miss the cues and clues that in our own culture—in an instant—would tell us what is happening.[22] To overcome this cultural mind-blindness, it is essential that we build the intercultural skills that widen our emotional radar and other sensory receptors and pick up those clues and awarenesses we would otherwise miss.

Intercultural Conflict Styles—The DEAD Model

Mitchell Hammer defined conflict style as interactional behavior that "reflects specific...patterns or tendencies for dealing with disagreements across a variety of situations."[23] To offset the effects of mind-blindness, recognize cultural differences, and help us read the dynamics of cross-cultural conflict situations, Hammer has devised an easy to understand framework that identify differences in conflict style when negotiating across cultures. An award-winning author[24] and researcher in the field of crisis mediation and conflict resolution, Hammer's Intercultural Conflict Styles (ICS-DEAD) framework looks at cross-cultural conflict from a culturally generalizable perspective. The ICS-DEAD describes four general cultural patterns and four conflict styles. Hammer begins by describing the four general patterns: the Direct and Indirect and the Emotionally Restrained and the Emotionally Expressive Cultural Patterns.

Direct Cultural Patterns

Hammer explains that cultures with a more direct communication style tend to frame their arguments and problem-solving language directly and precisely. This helps all parties to understand the issues and interests at play in a negotiation. According to Hammer, each party is responsible for verbalizing its "own concerns and perspectives and to verbally confront misperceptions and misunderstandings that can arise."[25] Such cultures, according to Hammer, tend to be comfortable with face-to-face negotiations that allow both sides to uncover misunderstandings, air grievances, and iron out disagreements. These cultures are also more likely to value those who can "tell it like it is" in ways that are both effective and appropriate. Good negotiators in these cultures are able to assert their needs or those of their group while maintaining some degree of politeness and tact. Hammer also described negotiations in these cultures to be typically characterized by appeals to reason based on facts or statistics. When problem-solving, they tend to "cut to the chase" and more often than not will focus on the solution rather than relationships or process issues.[26] This conflict style fits comfortably on the low-context communication side of Edward T. Hall's low-high context continuum.[27]

Indirect Cultural Patterns

Unlike cultures that are more direct, cultures that favor indirect communication patterns align more closely with the high context end of Hall's continuum.[28] Hammer describes these cultures as being tuned in to contextual messages that communicate outside the realm of the spoken word.[29] In negotiation or conflict situations, he asserts that verbal messages are intended more for the satisfaction of social expectations than to communicate interests or needs. When engaged in a dispute, cultures with an indirect style tend to view direct communication between parties as having a strong potential for making matters worse. Another difference is a tendency to use a more indirect means of persuasion. Instead of appealing to reason, indirect cultures tend to concentrate on facework. Ting-Toomey and Oetzel defined facework as the willingness and ability to "listen to the other person, respect the feelings of the other, and share personal viewpoints."[30] The importance of facework is evident in a preference for using third party intermediaries to settle disputes. Use of a trusted go-between allows all parties to save face while the mediator works to repair relationships and reach a resolution at the same time. In contrast to the direct style pattern of zeroing in on a resolution, the indirect cultural pattern is to approach problem-solving or conflict by focusing on repairing relationships. The solution is continually adjusted through the work of a third party until an acceptable resolution is reached.

When working with the ICS-DEAD model, negotiating teams should not become so focused on conflict style, that they forget the importance of enumerating interests and important facts and figures. These are important to the negotiation process regardless of the cultural context. However, the ICS-DEAD model can provide helpful insight into how the data can be effectively introduced into the process.

Emotionally Expressive Cultural Patterns

In emotionally expressive cultures, displays of emotion during a conflict tend to be expected and also valued. In these cultures, ventilating is generally accepted as a way to externalize or let out emotion.[31] In fact, the failure to externalize emotion in highly charged situations is often viewed with suspicion. Advising others to "relax" or "take it easy" is generally not positively construed and can be perceived as insincerity. In some emotionally expressive cultures, humor can be an acceptable way to reduce tensions.

Emotionally Restrained Cultural Patterns

In contrast to emotionally expressive cultures are the emotionally restrained patterns. In these cultures, strong feelings tend to be suppressed even when a person is greatly upset. Unlike expressive cultures, people from a restrained cultural background are apt to take a dim view of any attempt at humor in an emotionally charged situation. Emotions, of course, are enacted, but are more likely to emerge nonverbally and with minimal display. By allowing a glimpse of the underlying passion and commitment seething below the surface, these relatively subdued expressions of feeling can serve as an effective communication strategy when dealing with others comfortable with this pattern. Maintenance of a calm demeanor in the face of danger or high emotion tends to be highly prized. Consider these lines from Kipling's poem, *If*:

> *If you can keep your head when all about you*
> *Are losing theirs and blaming it on you,*
> *...you'll be a man my son!*[32]

The DEAD Conflict Styles[33]

Hammer's Intercultural Conflict Styles-DEAD Model identifies four distinct styles of cross-cultural conflict resolution. The four conflict resolution styles are: (a) Discussion, (b) Engagement, (c), Accommodation, and (d) Dynamic. As described above, the four styles are further sorted into four larger groupings of cultural patterns: (i) Direct and (ii) Indirect, and (iii) Emotionally Restrained and (iv) Emotionally Expressive. The chart in Figure 1 shows how the four cultural patterns intersect with the four conflict resolution styles.

The *Discussion, Engagement, Accommodation, and Dynamic Conflict Styles* form the ominous, yet oddly appropriate acronym *DEAD*. If one pays only scant attention to differing communication patterns and styles of conflict resolution, talks are more likely to end up "dead in the water."

Because the other parties in a negotiation cannot be counted on to be sensitive to our own preferences, it is doubly important that we understand how they handle conflict and negotiation. Such knowledge gives us a powerful negotiating tool.

Discussion Style. As the word discussion implies, people comfortable with this style prefer to talk through problems, positions, issues, and interests. The Discussion style is direct, but calm. "Say what you mean and mean what you say," is an American saying that describes this style. Facts and figures presented in a logical format are strong persuaders for individuals using this conflict style. Remaining calm while clearly describing issues, positions, and interests is the hallmark of this style. Proponents of this style believe that discussion reduces the possibility of misunderstanding while a "businesslike" atmosphere keeps everyone focused on issues and not personalities. The Discussion style aims for an expeditious completion of the negotiation. Unfortunately, this method for enhancing understanding is most effective when working with those who favor the same style. Negotiators from cultures where other styles predominate may find a Discussion-style negotiator either too direct or overly cold and calculating. They may feel that relationships are sacrificed just so the talks can proceed quickly. This style should seem familiar to most readers. It is the conflict and negotiation style that predominates in the U.S.

Engagement Style. Like the Discussion style, Engagement also has a preference for verbal directness in a negotiation or conflict situation. These two styles diverge in the way they handle displays of emotion. The Engagement style is direct and emotional. We might describe people who are comfortable with the Engagement style as "wearing their hearts on their sleeves." They are comfortable sharing their feelings, showing both commitment and sincerity.[34] Engagement-style negotiations tend to be animated and highly emotional when compared with Discussion-style interactions. Displays of emotion by Engagement-style negotiators can make their Discussion-style counterparts uncomfortable. On the other hand, anyone comfortable with an Engagement style may read the Discussion style demeanor as insincere or unwilling to acknowledge or engage with the intense feelings generated by the conflict or negotiation.

	Emotionally Restrained	Emotionally Expressive
Direct	Discussion Conflict Style • Direct • Emotionally Restrained	EngagementConflict Style • Direct • Emotionally Expressive
Indirect	Accommodation Conflict Style • Indirect • Emotionally Restrained	Dynamic Conflict Style • Indirect • Emotionally Expressive

Figure 1. Intercultural Conflict Styles (ICS), also known as "The DEAD Model."

Accomodation Style. The Accommodation style, like the Discussion style, is emotionally restrained, but people preferring this style tend to be indirect in the way they approach conflict resolution. This style relies on context, ambiguity, metaphor, and third party intervention to improve any verbal confrontations between parties.[35] Relational harmony is typically maintained by hiding one's emotional discomfort. Those who are comfortable with this style are adept at reading ambiguous high context messages. As previously stated, use of third party intermediaries are common. In discussing conflict resolution in Korea, the late L. Robert Kohls, a cross-cultural training pioneer, suggests one should locate a go-between earlier on in the process than you would in the U.S. According to Kohls, "the use of mediators is common in Korea and does not imply the extremity of conflict it does in the United States."[36] A person accustomed to a direct style is likely to suffer from mind-blindness and may be unaware that a problem even exists. In such cases, a conflict may suddenly burst forth "like a volcano exploding."[37] When the bewildered American asks the aggrieved parties what happened, they are likely to say, "We WERE telling you very loudly," but not in words.

Dynamic Style. The last style in the ICS-DEAD framework is Dynamic. Like the Accommodation style, Dynamic negotiators and disputants tend to use indirect messages to settle disagreements, but with a more emotionally intense verbal style. Hammer explains that the Dynamic style is marked by "strategic hyperbole, repetition of one's position, ambiguity, stories, metaphors, and humor along with greater reliance on third party intermediaries."[38] Dynamic negotiators are accustomed to working with intermediaries and are quite at home with displays of anger or emotion. As indirect communicators, they are likely to describe themselves as good observers of behavior capable of providing helpful solutions to all parties in a dispute.

Discussion-style negotiators may view a Dynamic-style counterpart as an overly emotional person who rarely gets to the point. To discern underlying values, arranging for a third-party intermediary or go-between may be the best way to uncover the underlying causes while still maintaining the relationship. As the Arab proverb tells us, "It is good to know the truth, but it is better to speak of palm trees."[39] A person operating from the Dynamic style may need to overcome negative feelings that the Discussion style counterpart is insincere, insensitive, and impatient.

Cultural Differences in Conflict Style: The ICS-DEAD Model is the product of a comprehensive research project conducted by Hammer over a period of several years. His findings were formally published in an academic journal in 2005.[40] The data was drawn from a 106-item survey. The survey questions were gleaned from a broad review of the literature related to cross-cultural communication. Hammer administered the survey to 510 culturally diverse respondents. As a final step, Hammer evaluated those findings, re-worked the survey into a more user-friendly format, and administered the revised version to a new sample of 487 respondents from diverse cultural backgrounds.[41] Both surveys produced results that proved to be both statistically reliable and valid. Hammer cautions that "all cultural patterns exist in all cultures—but some are preferred more than others," depending on the culture.[42] According to his findings, a wide variety of communication and conflict styles are employed on every continent. The following list is arranged alphabetically and contains examples from each continent and region:

- Africa: Three styles predominate: *Engagement Style* (West Africa, e.g., Nigeria), *Accommodation Style* (Horn of Africa, e.g., Somalia), and *Dynamic Style* (The Maghreb of North Africa and Egypt)
- Asia/Pacific: Four styles predominate: *Accommodation Style* (East Asia, e.g., Japan; Southeast Asia, e.g., Cambodia), *Dynamic Style* (Indian Subcontinent, e.g., Pakistan), *Discussion Style* (Indian Subcontinent, e.g., India; Pacific, e.g., New Zealand), and *Engagement Style* (Former Soviet Union, e.g., Russia)
- Europe: Two styles predominate: The *Discussion Style* (Northern Europe, e.g., Germany) and *Engagement Style* (Southern Europe, e.g., France; Eastern Europe—Former Soviet Union, e.g., Ukraine and Belarus)
- Latin America and the Caribbean: Two styles predominate: The *Accommodation Style* (e.g., Mexico) and *Engagement Style* (e.g., Cuba)
- The Middle East: Two styles predominate: *The Dynamic Style* (e.g., Iraq) and *Engagement Style* (e.g., Israel)
- North America: One style predominates: *The Discussion Style* (The U.S. and Canada)

Openness to differing communication, negotiation, and conflict styles leads to understanding. The ability to remain open and suspend judgment is the key to effective leadership of coalition teams and can provide insight into cultural differences when engaged in a cross-cultural negotiation. Because Discussion is the predominant conflict and negotiation style in North America, American supervisors or negotiators are likely to misunderstand or underestimate those with differing approaches unless they have developed a clear understanding of the meaning behind the behavior.

A Final Note about the ICS Model: DEAD

This conceptual GPS, while extremely useful in its ability to increase your awareness and understanding of cultural differences, cannot replace mindful and reflective communication practices on your part. Cultures are not either Discussion Style or Engagement Style. They do not have either Indirect Cultural Patterns or Direct Cultural Patterns. Cultures are extremely complex totalities rife with paradoxes and contradictions. Cultures are never "either/or." They are always "both/and." With these caveats in mind, remember that all the cultural patterns described in this model can be found in all cultures. However, the research on which the model is based has shown that some styles are preferred more in some cultures than in others.[43]

A WORD ABOUT *PRIO*: PATIENCE, RESPECT, INTEREST, & OPENNESS

You have probably noticed that PRIO, the affective skills of *Patience, Respect, Interest, and Openness*, have been mentioned individually throughout this article as keys to thoughtful negotiation and conflict resolution. [These skills] will serve you well when communicating cross-culturally.

Patience. Suffice it to say that any communication taking place across cultures requires patience. Negotiations and conflict resolutions will always take longer when enacted across cultures. If you reflect back on the differing cultural patterns and conflict styles you have just read about, you will notice that most cultures need more time to come to an agreement than North Americans and Northern Europeans, and that is without taking into consideration language differences and other cultural impediments to communication. Plan your cross-cultural negotiations to allow enough time to accommodate cultural differences. Your negotiation may not take thirteen days of nearly round-the-clock discussions like the Camp David Accords, but time is important and must be factored into your plan.

Respect. You may not always respect those with whom you have to do business, but in an intercultural setting, you must show respect for the culture if you expect an equitable resolution. Respect goes hand in hand with patience. One way of showing respect is by taking time to learn what cultural values are entwined with the interests of the other party. As Americans we tend to value the product or solution to the problem more than the process. If we learn to also respect the process, we may be more likely to get the product or solution we seek. Respecting the process means being sensitive to relationships, utilizing third parties when necessary, and understanding the meaning behind emotional expressiveness when it emerges.

Interest. Interest, as mentioned earlier in this article, requires that you find out as much about the other side's position, issues, interests, and values. When negotiating cross-culturally, seek out knowledgeable experts on the culture. Be sure to talk to host nationals as well as Americans. And, if needed, do not hesitate to locate a trusted third party to help you and the other parties concerned. He or she can provide valuable insight into the process.

Openness. Reading this article should have raised your awareness of cultural differences and should also help you to remember to suspend judgment until all the facts are in. This skill requires time, effort, and practice. But awareness is the first step and that step will lead to the development of an open attitude. Openness is the key to learning about cultures independently and how to navigate them appropriately and effectively. By cultures, I refer not only to other cultures, but your own as well. A better understanding of cultures in general will lead you to a better understanding of yourself and the world around you.

CONCLUSION

Effective cross-cultural negotiation and conflict resolution has certain requirements. You must do your homework. Understand your own position, interests, and values as well as those of all the other parties involved. Try to discern when values are involved as well as interests. Interests are important, but they are not the most important consideration. As Forester stated, "values run even deeper than interests," and this is true no matter the context or location of the interaction.[44] Sometimes our negotiating partners may place a much higher value on face or respect than on material gain. You also need to plan ahead. Indeed, planning is the single most important element in preparing for a negotiation. Yet great planning will not help in cross-cultural negotiations if you have not visited the [culture] and prepared yourself to handle behavioral and institutional differences, and discerned their linkages to cultural values, beliefs, and assumptions.

It's also important to remember that as human beings we have trained ourselves to thin-slice in every interaction, but we lose our adeptness and become mind-blind as soon as we cross the cultural Rubicon. Understanding the ICS-DEAD Model can help us as we cross that river. However, unless we utilize the affective PRIO skills, our attempts at effective interaction may founder on the shoals of ineffective communication.

Good planning, active listening, and a mindful approach to any conflict resolution can often produce unexpectedly positive results. May your cross-cultural journeys be free of conflict. It is often the need to settle issues and solve problems that helps us to build those relational bridges that serve the greater strategic mission.

END NOTES

1 Jeanne M. Brett, Negotiating Globally: How to Negotiate Deals, Resolve Disputes, and Make Decisions Across Cultural Boundaries, 2nd Ed. (San Francisco, CA: John Wiley & Sons, Inc., 2007), 1.

2 Ibid, p. 3.

3 Brett, Negotiating Globally, 10.

4 Roger Fisher and William Ury, Getting to Yes, 2nd Ed. (New York: Penguin Books, 1991), 41-42.

8 Ibid, 42.

6 Jimmy Carter, Keeping Faith: Memoirs of a President. (New York: Bantam Books, 1982).

7 Ibid, 416.

8 Fisher and Ury.

9 Don Conlon, "Successful Negotiation: Creating Value Through Collaboration" (Eli Broad School of Management, Michigan State University, East Lansing, MI, notes), 7.

10 A stereotype is the practice of attributing a personal characteristic, behavior, value, or belief to a group.

11 Brett, 10.

12 Stella Ting-Toomey and Leeva Chung. Understanding Intercultural Communication. (Roxbury Publishing Co., 2005), 283.

13 Brett, 10.

14 John Forester, "Dealing with Deep Value Differences," in Lawrence Susskind, S. McKearnan, and J. Thoman-Larmer (Eds.), The Consensus Building Handbook: Comprehensive Guide to Reaching Agreement. (Thousand Oaks, CA: Sage Publications, Inc., 1999), 463.

15 Ibid.

16 Carter, 335.

17 Jimmy Carter, Keeping Faith: Memoirs of a President. (New York: Bantam Books, 1982), 416

18 Stella Ting-Toomey and John Oetzel, Managing Intercultural Conflict Effectively. (Thousand Oaks, CA: Sage Publications, Inc., 2001), 17.

19 Malcolm Gladwell, Blink: The Power of Thinking without Thinking. (New York: Little, Brown, and Co., 2005), 23.

20 Ibid.

21 Ibid, 214.

22 Ibid, 221.

23 Mitchell R. Hammer, The Intercultural Conflict Style Inventory: ICS Interpretive Guide. (Berlin, MD: Hammer Consulting, 2003), 6.

24 Hammer's book, Saving Lives: The S.A.F.E. Model for Resolving Hostage and Crisis Incidents, won the Outstanding Scholarly Book Award for 2008.

25 Mitchell R. Hammer, "Chapter 10: Solving Problems and Resolving Conflict Using the Intercultural Conflict Style Model," in Michael A. Moodian (Ed.), Contemporary Leadership and Intercultural Competence: Understanding and Utilizing Cultural Diversity to Build Successful Organizations. (Thousand Oaks, CA: Sage Publications, Inc., 2009), 224

26 Ibid, 225.

27 Edward T Hall. Beyond Culture, (Garden City, NY: Anchor Books, Doubleday, 1977), 105-106.

28 Ibid.

29 Ibid.

30 Ting-Toomey and Oetzel, Managing Intercultural Conflict Effectively, 50.

31 Hammer, Chapter 10 in Moodian, 225.

32 Rudyard Kipling, If: A Father's Advice to His Son, (New York: Macmillan and Company, Ltd., 1917), 1.

33 Hammer, Chapter 10 in Moodian, 226-227.

34 Ibid," 226.

35 Hammer, Chapter 10 in Moodian, 227.

36 L. Robert Kohls, Learning to Think Korean: A Guide to Living and Working in Korea, (Boston, MA: The Intercultural Press, 2001), 175.

37 John C. Condon, With Respect to the Japanese: A Guide for Americans. (Yarmouth, ME: The Intercultural Press, 1984), 45.

38 Hammer, Chapter 10 in Moodian, 227

39 Hammer, ICS Interpretive Guide, 9.

40 Mitchell R. Hammer, "The Intercultural Conflict Style Inventory: A Conceptual Framework and Measure of Intercultural Conflict Resolution Approaches," International Journal of Intercultural Relations 29 (2005): 675-695.

41 Ibid, 688.

42 Ibid, 14.

43 Ibid, 15.

44 Forester, "Deep Value Differences," 463.

ABOUT THE AUTHOR

Dr. John W. Miller is professor of Culture and Language Studies at the US Air Force's Squadron Officer College.

This article is used with permission of the author.

16.4 Preventive Diplomacy: Training a New Generation for Peace

By Carl Hobert

OBJECTIVES:

15. Define the term "preventive diplomacy."
16. Describe the differences between Track One, Track Two, and Track Three diplomacy.
17. Describe the core principles of preventive diplomacy.
18. Compare "positional bargaining" with "principled" or "integrative bargaining."

"We are caught in an inescapable network of mutuality, tied in a single garment of destiny."

Martin Luther King, Jr.,
"Letter from Birmingham Jail"

Beginning in 1983, I devoted years of my life as an independent school instructor to teaching international conflict resolution, before I decided that there is no such thing—at least not in the way we tend to imagine it. In the global arena particularly—where the causes of border skirmishes, assassinations, acts of terrorism, coups d'état, or all-out warfare have such deep roots in historic, religious, political, economic, and social inequities—resolving a conflict often doesn't make it go away forever. As the daily news headlines from myriad global hot spots remind us, as long as the root causes of a conflict linger, or memories of it have yet to heal, the potential for divergence, discord, tensions, clashes, or renewed all-out conflict remains real.

As a direct result of the war- and conflict-riddled world in which our students are coming of age, I find it more helpful than ever to talk with them not about conflict resolution, but rather about conflict management and prevention, through the art of negotiation and the principles of preventive diplomacy. Young people take to preventive diplomacy naturally, even eagerly. Most children are old hands at conflict and negotiation at a personal level with parents, siblings, teachers, and peers. Some in the U.S., and even more elsewhere around the globe, have witnessed much worse, too: parents, siblings, teachers, and/or friends killed in armed conflicts, communities and whole cultures devastated by violence. In some places, children themselves are often the well-armed killers, trained by adults to do their bidding. Whatever their proximity to violence, whether they see it on television, or breathe it or feel the threat of it right in front of them 24/7, children may feel called to peace—or called to a "piece of the action" of bloodshed, of vengeance. What is clear to me now is that schools can and should play a role in helping young people—our future negotiators—learn the tools of preventive diplomacy. In this increasingly interconnected world, such knowledge may be one of our best hopes for tangible peace, today and in the future.

Preventive diplomacy has had a long and instrumental role in international relations. World leaders and foreign policy experts have recognized it as one of the most powerful alternatives to armed conflict, and essential if we are to prevent globally catastrophic wars and other forms of violence. Former UN Secretary-General Boutros Boutros-Ghali described preventive diplomacy as "diplomatic action to prevent existing disputes from arising between parties, to prevent these disputes from escalating into conflicts, and to limit the spread of the latter when they occurred." In the field of preventive diplomacy, two distinct veins have emerged: Track One and Track Two diplomacy. What are these? "Track One diplomacy" refers to ongoing, formal negotiations between official representatives of nation-states—such as presidents, prime ministers, foreign ministers, and/or ambassadors—to resolve or prevent conflicts. "Track Two diplomacy" refers to more subtle social assistance by professional, nongovernmental organizations (NGO's) or persons—i.e., appointed arbitrators or organizations such as Doctors Without Borders—to ease tensions between nation-states. These non-military, Track One and Track Two diplomatic strategies have been helpful to some extent in addressing potential crises between nations or peoples before they erupt again in violence in such powder-keg areas as Northern Ireland, the Indo-Pakistani Kashmir region, and Bosnia.

In addition to these well-established forms of preventive diplomacy, I believe there to be another, equally—if not more—valuable form of international conflict prevention: Track Three diplomacy. This form of preventive diplomacy—which we employ in Axis of Hope, an educational organization that I founded in 2002—involves creative educational efforts to teach conflict analysis, management, and prevention to students around the globe. These efforts help to deepen students' understanding of the religious, cultural, socioeconomic, and psychological roots of geopolitical conflicts, and to provide them with the tools required to help bring more peaceful coexistence to these areas of conflict. How do we teach students Track Three diplomacy? In Axis of Hope, we do it by transporting them intellectually from the familiar territory of their schools (riddled as they are with their own emotional minefields) to a more challenging, distant culture in crisis: the Middle East. For one-half day to five days, middle and/or upper school boys and girls with whom we work take on the roles of Israeli or Palestinian moderates or extremists, members of a Track One diplomatic quartet, or people employed by the Track Two World Bank—roles they play based on the case study of the Arab-Israeli conflict that we authored, entitled "Whose Jerusalem?" a Harvard Business School-type case study on the Middle East conflict.

During the seminars, we begin by having students read the assigned case-study history of the Arab-Israeli conflict that details the religious, social, cultural, and economic factors integral to the analysis of the conflict and that offers an in-depth chronology of the conflict. We also offer lectures on how to analyze the conflict from a negotiator's point of view and how to effectively practice the art of negotiation. Perhaps more importantly, students participate in "intellectual outward bound" role-play exercises representing the aforementioned and other stakeholders on all sides of the conflict. By the end of the negotiating exercises, students learn valuable lessons about how they might promote peaceful coexistence in the Middle East, and how they might relate the lessons they have learned to more successful coexistence efforts right here in the U.S., in their own schools, and in their own homes.

These pedagogical efforts provide students with a progressive form of learning in which they can hone their diplomatic skills in the safe space of an educational environment—allowing them to take risks, make mistakes, and live to tell about it. All of these efforts are based on four key points, which we call "The Preventive Diplomacy Core Principles."

THE PREVENTIVE DIPLOMACY CORE PRINCIPLE

Classic negotiation and conflict resolution often eschews the "I win, you lose" negotiation style, also described as "positional bargaining" in which "hard" bargainers will do anything to win and "soft" bargainers will give up the ship to preserve the relationship with the other side's representative. Neither leads to a fair, sustainable conclusion. Preventive diplomacy training for students relies on principles and practices adapted from the work of many in the field of conflict resolution and negotiation whose insights now define approaches used around the globe in business, government, personal relationships, and other arenas. While our key concepts come from a variety of sources, the most important ideas in our teaching come from the book Getting to YES: Negotiating Agreement Without Giving In, by Roger Fisher, William Ury, and Bruce Patton, of the Harvard Negotiation Project. In creating preventive diplomacy principles and practices for students, we've drawn extensively from their straightforward method for negotiation and conflict management's four basic principles: (1) focus on interests, not positions, (2) separate the people from the problem, (3) invent options for mutual gain, and (4) learn how to talk so people will listen. In our experience, these key principles quickly engage students and turn a complex subject (for example, the Arab-Israeli conflict) into an effective, hands-on learning experience.

FIRST:
FOCUS ON INTERESTS, NOT POSITIONS.

For the purpose of teaching students useful conflict analysis, management, and prevention skills, the first pillar is "principled" or "integrative" bargaining, in which the negotiating parties focus on reconciling their interests rather than their positions or differences. Understanding the other side's interests gives more precise meaning to the problem.

Awareness of the fact that the most prevailing interests are most often very basic human needs is vital, too. These basic needs include power, security, a sense of belonging, and recognition. After both—or all—sides' interests are clearly defined, it is then up to the negotiating sides to find shared interests, as well as conflicting ones, because underneath differing positions there can also be subtle, shared, compatible interests between and among enemies. Although Palestinians and Israelis—or students playing the roles of these key stakeholders in the Middle East conflict—may not believe in the same faith, all of the negotiators have families, friends, personal interests, and amazing personal stories of love and loss. Students must learn to study the person or persons with whom they will be negotiating, making an effort to understand their shared personal interests—as well as how to make their, and their adversaries', interests "come alive" in negotiations. The savvy student negotiator learns how to discuss these shared and conflicting interests in creative, energetic ways, and how to bargain in concrete but flexible ways. Establishing a "common interest" focus from the outset of negotiations leads to more collaborative discussion, a better synthesis of ideas, and potentially innovative solutions for problems that previously appeared intractable.

The person with whom he or she is negotiating does not just possess the thoughts, the ideas, and the official positions of the other side's government—or the other side's grade level or sports team or social network. He or she also possesses many of the thoughts, ideas, positions, and interests that the other negotiator deems close to his or her heart as well. If a negotiator is able to smile and focus on these common interests first, instead of always focusing on conflicting ideas and frowning and arguing and walking away, he or she gains much more respect from the other side from the outset, and in the long run. As Fisher says: "Behind opposed positions lie conflicting interests, as well as shared and compatible ones."

SECOND: SEPARATE THE PEOPLE FROM THE PROBLEM.

Preventive diplomacy teaches students what Fisher, Ury, and Patton taught their students: "Don't be hard on the other side." To be precise, they urge us to "be hard on the problem, but be easy on the people," if we hope to negotiate successfully. Negotiators are, after all, people first. First, students learn to build a working relationship with the negotiator representing the other side. Then, they learn to tackle the problem. In doing so, they are taught to imagine why the other side's representatives are arguing their case the way they are. The talented negotiator first separates the people he or she is working with from the problem they are discussing.

The next vital step is being able to "walk in the shoes of the other side." One handy example: before using the "Whose Jerusalem?" case study as a role-play exercise, we have teachers ask students, well before the activity begins, to identify which sides they want to represent. For example, do they want to represent Likud (Israeli right wing, or conservative party members) or Hamas (the Palestinian extremist organization, with known political and terrorist wings)? If a student indicates that he or she would like to be a Likud representative, the teacher can surprise the student by assigning him or her to play the role of the opposite position, or Hamas, requiring this student to learn to understand, and then defend, the other side. We have found that this not only allows students to learn more about all sides of a conflict, but it also helps them to be more compassionate when arguing in favor of their original position at a later date. They tend to listen more carefully to all sides, acknowledge what is being said, speak more effectively in order to be understood, and learn the importance of the old diplomatic term: "We agree to disagree." In short, students learn that a vital diplomatic skill is to research and understand all sides in a conflict. This leads to quicker, more effective negotiations and problem solving in the long run.

THIRD: INVENT OPTIONS FOR MUTUAL GAINS.

Negotiators often offer little, demand much, and stubbornly haggle over a single quantifiable issue like money, as if they are in a bazaar trying to talk a merchant down. The good negotiator creates what Fisher, Ury, and Patton call "mutual gains" in negotiations, so that negotiators on both sides are able to achieve some—if not all—of their goals together, without compromising the interests of their own constituents.

When negotiating, students learn never to assume that there is only one answer to a question, or one way to solve a problem, or one outcome a negotiator must seek. They learn how to enter negotiations in a very open-minded way, with an ability to invent multiple options for outcomes. We teach them to listen to the outcome options or back-up plans of the other side, too. A talented negotiator will always prioritize and preview desired outcomes, invent alternatives if needed, and develop a step-by-step plan to achieve them, in a process that involves the other side's negotiator(s). We teach students to not only think about solving their problem, but to help the other side solve its problems as well. Identifying myriad interests that both sides share—and inventing options that could satisfy both parties—is crucial. This creative, inventive brainstorming process of developing multiple options is vital to achieving mutual gains.

It may be difficult to think of students agreeing to "lasting peace" in the Middle East at the end of a role-play exercise. But what if you up the ante by setting a time limit, giving students only one night, or one hour after a day of conflict-management exercises, to, say, write a letter to former British Prime Minister Tony Blair or to Secretary of State Condoleezza Rice before he or she is to depart for the Middle East on a peace-seeking trip, outlining ideas about creating peace to the Middle East? In such a pressure scenario, students learn to "rally"—to invent broader and more creative options for mutual gains in a conflict. Students learn how to no longer simply represent the Israeli right wing or Hamas, but work in small groups at new negotiation tables marked "Education," "Health Services," "Defense," "Politics," and more. Here, they learn to identify shared interests and negotiate in a different environment, where new ideas—and new, creative options, rather than simply parties' interests—are being discussed. This is known as "diplomatic brainstorming for the win-win," during which time students search for new ways to create mutually agreed upon solutions in these different areas.

FOURTH: LEARN HOW TO TALK, SO PEOPLE WILL LISTEN.

It is essential that negotiations produce agreements amicably and efficiently. Use of proper body language, the appropriate choice of words, and the correct tone of the voice are crucial diplomatic tools students learn to refine before going to the negotiating table. The good negotiator is one who is able to establish easy two-way communications, so that his or her negotiating relationship is, from the outset, not adversarial. We teach students to build a good, side-by-side working relationship. We often ask students guiding questions. "Are you seated in a chair during negotiations, or are you standing beside the chair—or on the chair, or on top of the table—trying to show superiority? While negotiating, are you screaming or raising your voice, or are you negotiating with a firm yet respectful tone? Are you speaking in an arrogant manner, or in a humble way? Are you leaning back in your chair and crossing your arms and legs, removing yourself physically from the talks, or are you leaning forward and with arms opened, interested in and open to the negotiation procedure? And, finally, we teach students that the word "silent" spelled another way is "listen." We ask: "Do you show respect to the other side in negotiations by remaining silent and listening often?"

Peace is a process, not a prize. There is no such thing as "lasting peace." In international conflict, peace isn't something we achieve and then leave behind, assuming that a peace accord or a treaty is part of a completed task, never to be revisited. We now know that what matters is international conflict management, achieved through ongoing preventive diplomacy, including constant educational exercises in conflict analysis, management, and prevention. As future leaders, our students can learn to see peace as an architectural process that must be discussed and negotiated and drafted together, and refined over and over again—before it is even built in the form of a temporary peace treaty. And, then, as I always tell students, days or months or years later, this beautifully crafted peace model must be remodeled again.

By teaching future leaders to develop trust, compassion and empathy for one another, and for people around the world, educators can help change the landscape of conflict and help create the prospect of future peace. U.S. independent schools are doing an excellent job of focusing students on global issues, but they might contemplate taking the next step in helping students learn how to deal with these complex issues in a hands-on way. Allowing students to participate in open, honest discussions of thorny world issues will teach them essential preventive diplomacy skills that will last a lifetime.

ABOUT THE AUTHOR

Carl F. Hobert, a graduate of the Tufts University Fletcher School of Law and Diplomacy, is executive director of Axis of Hope and is completing a book entitled Axis of Hope: Teaching Global Responsibility in U.S. Schools (Beacon Press). He may be reached at carl.hobert@axisofhope.org.

From: Carl Hobert, "Preventive Diplomacy: Training a New Generation for Peace," *Independent School Magazine* (Spring 2008). Used with permission.

16.5 The Not-So-Black Art of Public Diplomacy

By Humphrey Taylor

OBJECTIVES:

19. Define the term "spin."
20. Define the term "public diplomacy."
21. Describe the difference between the use of hard power and soft power for diplomacy.
22. Identify tools that a nation can use to influence public opinion.

How is it that the country that invented Hollywood and Madison Avenue has such trouble promoting a positive image of itself overseas?

—Rep. Henry Hyde, October 2001

National leaders have the power to shape foreigners' opinions of their countries, for better and worse. This is true, of course, for such giants as FDR, Churchill, de Gaulle, Hitler, Stalin, and Mao Tse Tung; so too Bush, Blair, Merkel, Chirac, Sarkozy, and Putin have all changed the way foreigners see their countries. Their influence is a result of many factors, including substance, style, and spin. Substance relates to policies, and in particular their foreign policies. Style is about charisma and personal chemistry; here President John F. Kennedy, who was wildly popular abroad, comes to mind. Spin is a pejorative for a legitimate function, communication—how leaders and countries explain themselves and their policies to the world. In recent years, a new phrase has sometimes been used to describe these communications: public diplomacy.

The poet Robert Burns, in his "Ode to a Louse," wrote: "Oh would some power the giftie gie us/to see ourselves as others see us./ It would from many a blunder free us, and foolish notion." Unfortunately, it is probably true that most people in most countries do not see themselves as others see them. History books almost everywhere tend to teach children that their country and their people are better than others, and the media and politicians pander to these beliefs and prejudices. This is true not just of strong and powerful countries but of small countries and even tribes. Serbs, Bosnians, Albanians, and Croats all have very different history books and are shocked that the rest of the world does not share their view of history. While objective histories see most Balkan peoples as both the perpetrators and victims of atrocities, each group usually sees themselves only as victims with many reasons to feel proud of their history and no reasons to feel ashamed.

My mother was born in England in 1894, at the apex of British imperial self-confidence and pride. When still young, she was stunned to meet a young French boy who told her he was proud to be French. How she wondered, could anyone be proud to be French, or any nationality other than British? It was incomprehensible to her. Everyone, she assumed, knew that Britain was the best country in the world.

Similarly, some Americans see themselves as latter-day Athenians, the defenders of a great democracy pitted against ruthless and undemocratic Spartans. Sometimes this may be a useful analogy. However, others see Americans as the ruthless Athenians who crushed the neutral island of Melos, killing the men and enslaving the women and children. In Thucydides' famous account, the Athenians demanded that the Melians surrender because Athens was much stronger than Melos and that:

You know as well as we do that, when these matters are discussed by practical people, the standard of justice depends on the quality of power to compel and that in fact the strong do what they have the power to do and the weak accept what they have to accept.

One need not look hard to see shades of "you are either with us or against us," which has sometimes appeared to be the position of the American government under the administration of President George W. Bush.

THE IMPACT OF IRAQ

The impact of the war in Iraq on world opinion has, of course, been overwhelming. As early as 2003, under the headline "Foreign Views of United States Darken after September 11," Richard Bernstein wrote in *The New York Times* that:

> The war in Iraq has had a major impact on public opinion, which has moved generally from post-9/11 sympathy to post-Iraq antipathy, or at least to disappointment over what is seen as the sole superpower's inclination to act preemptively, without either persuasive reasons or United Nations approval.

To some degree, the resentment is centered on the person of President Bush, who is seen by many of those interviewed, at best, as an ineffective spokesman for American interests and, at worst, as a gun slinging cowboy knocking over international treaties and bent on controlling the world's oil, if not the entire world.

This negativity was highlighted in an August 3, 2006, column in the *Financial Times* by a distinguished former British diplomat, Rodric Braithwaite, calling for the resignation of Tony Blair. At the time, Blair, the staunchest ally of President Bush, had the lowest poll ratings of his three-term premiership. "Blair's total identification with the White House has destroyed his influence in Washington, Europe and the Middle East," Braithwaite wrote. "Who bothers with the monkey if he can go straight to the organ-grinder?" When Americans re-elected President Bush in 2004, the popular British tabloid, *The Daily Mirror*, filled its front page with the words "ARE THEY MAD?"

Another factor that has fueled hostile criticism is climate change—the unwillingness (until recently) to accept that this is a serious problem made worse by human activity, and the rejection of the Kyoto Treaty. This led to the isolation of the United States at the recent United Nations Conference on Global Warming in Bali. *The New York Times* report from Bali referred to "the escalating bitterness between the European Union and the United States," and the very strong criticism of U.S. policies by "countries rich and poor." At one point the audience booed the American delegate.

As the Bush presidency winds down, there is a new focus on what will constitute the president's foreign policy legacy. It will surely include his record in Iraq, Afghanistan, North Korea, perhaps the Israeli-Palestinian conflict, and reflect the pervasive issues of Guantanamo and climate change. It also seems likely that one element of his legacy abroad will be lost trust and respect, and more hostility and criticism.

In general, favorable views of the United States have fallen steeply over the last seven years—but possibly not so far as some critics and pessimists believe. The Pew Global Attitudes Project provides trend data between 1999/2000 and 2007 for 25 countries. At the beginning of this period, majorities in 22 countries had favorable attitudes to the United States. In 2007, 13 still did. But, in 1999/2000 more than 60 percent of the public in 13 countries had favorable views of the United States. However, in 2007, this was true in only six countries.

Some of the largest declines in favorable attitudes have occurred in countries we usually think of as allies and friends, with falls of 32% in Britain, 23% in France, 48% in Germany, 23% in Italy, 32% in the Czech Republic, 25% in Poland, 43% in Turkey, and 46% in Indonesia. (This survey also shows a huge increase in Nigeria with regard to trust in the US for which I can offer no explanation.)

Major drivers of this decline have, of course, been foreign policy, the war in Iraq, and the so-called war on terror. The Pew Global Attitudes Project provides trend data on attitudes to the U.S.-led war on terror for 31 countries between 2002 and 2007. In 2002, not long after the 9/11 attacks, majorities in 23 of these 31 countries supported the war on terror. By 2007, majorities in only 11 countries still did so. And, in countries with even more favorable views of U.S. policy, the drops were just as sharp: in 2002, more than 60 percent supported the war in 19 countries; in 2007 they did so in only three countries.

Of course, all of these numbers can be expected to change between now and President Bush's departure from the White House, but for now this aspect of his legacy looks bleak.

WHAT IS PUBLIC DIPLOMACY?

Joshua Fouts, director of the Center on Public Diplomacy at the University of Southern California's Annenberg School for Communication, defines public diplomacy as a "government reaching out to a public or polity to explain its cultures, values, policies, beliefs and, by association, to improve its relationship, image and reputation with that country."

The phrase "public diplomacy" is relatively new, as is the fact that the State Department employs an Undersecretary for Public Diplomacy. However, governments and leaders have engaged in public diplomacy in the past, even if they did not use the phrase. The Voice of America, Radio Free Europe, Radio Sawa, Radio Marti, and the activities of the U.S. Information Service and sometimes the Central Intelligence Agency (CIA) are all part of American public diplomacy. Arguably public diplomacy is a polite phrase for propaganda when the propagators are the good guys who, unlike Goebbels or Stalin, are only trying to tell the truth about world events. But who are the good guys? Sometimes that is in the eye of the beholder.

Before Pearl Harbor, Winston Churchill sought desperately to influence American opinion and win support for the Allies in World War II. Lord Halifax, the British ambassador in Washington, and Isaiah Berlin, who was working in the British Embassy, were charged with the task of competing with such isolationist figures as Charles Lindbergh and Father Coughlin for American hearts and minds. They cultivated opinion leaders and fed information to friends in the media. Since then many countries have paid public relations firms to tell their stories and promote their countries to the American people. More recently Israel, and its friends in the United States, along with other lobbies, have done a particularly effective job of promoting positive attitudes toward the country and its causes.

But if public diplomacy is not new, the focus on it has palpably increased. What has changed is the belief that the public relations techniques used domestically by politicians, corporations, and advocacy groups to influence the attitudes and perceptions of the American public can also be used by governments to influence public opinion in other countries. Madison Avenue and the public relations' industry know how to influence hearts and minds. Why not use their skills to win more friends around the world? Or to reduce Muslim hostility to the United States? The failures of public diplomats such as Charlotte Beers and Karen Hughes show how difficult this is. Nevertheless, many countries increasingly buy full-page ads and multi-page supplements in major newspapers and magazines to tell Americans how wonderful their countries are.

THE LIMITS OF SPIN

Underlying much of the political support for American public diplomacy is the belief that public relations techniques can make world opinion more supportive of, or at least less hostile to, U.S. policies—*without any change in these policies.* Some advocates seem to believe that, since American policies are inherently honorable and ethical, all that is needed is to explain them more effectively and people will think better of America. Corporate executives often feel they can improve their companies' reputations, and politicians their popularity—all through communication. Occasionally, but not often, they are right.

Even where press coverage of a country improves, it is difficult to determine how much of the improvement was caused by public diplomacy. An interesting column in *Izvestiya* (mentioned in *The Week,* August 18, 2006) reported: "To change world opinion, the Kremlin has turned to an American public relations firm. Several months ago, the Kremlin hired Ketchum, hoping to combat the 'almost entirely negative' press Russia was getting in the run-up to the Group of Eight conference in St. Petersburg." Ketchum used its "numerous connections in journalism to plant 'objective and even favorable' articles about Russia in newspapers in the U.S. and Britain. Still, whether those articles had any substantial effect on policymakers is debatable. Russia expert Marshall Goldman of Harvard says the reason Russia wasn't criticized at the summit was because everyone was distracted by the war in the Middle East. 'As far as I know,' he said, 'Ketchum had nothing to do with what was happening in Lebanon.'"

Sometimes, it may not be possible to separate public diplomacy from traditional diplomacy—to say where one ends and the other begins. One of the great successes of President George H. W. Bush's diplomacy in the first Gulf War was in forming a U.S.- led coalition that included Muslim and Arab forces. Almost all the world's governments, explicitly or implicitly, supported the liberation of Kuwait and the invasion of Iraq. One of the reasons for not "pushing on to Baghdad" was the fear of getting bogged down there. However, another important consideration was the belief that the coalition would fall apart and alienate both governments and publics in the Muslim world. This was a case in which an understanding of foreign public opinion influenced policy, and not merely an exercise in communication.

Effective public diplomacy should, I believe, work hand-in-glove with traditional diplomacy. It is understood that traditional diplomacy involves give and take, that compromises are often necessary, and that two-thirds of a loaf (or even half) is better than no loaf. Likewise, our public diplomacy should involve both give and take. It should help improve communications but it should also influence what the United States government does, and what our leaders say or do not say.

In the corporate world, wise chief executive officers (CEOs) make sure that their senior communications managers—who are the guardians of their companies' reputations— report directly to them. An effective approach to corporate public relations is not didactic: "This is what we are doing, put the best spin on it." It is interactive: "What should we do as a company and what should I do as the CEO—regarding actions, policies, programs, and communications—to ensure that this company and its products and services are liked and trusted by the public, our customers, employees, suppliers, legislators, regulators and shareholders?" Successful public relations directors do much more than just manage communications.

If traditional diplomacy often relies on "hard power," the use or possible use of military or economic strength to achieve its ends, public diplomacy often uses "soft power"—cultural, political, educational, and economic forces. Successful diplomacy based on hard power may cause people to respect, but also to fear, dislike, and distrust its users. Successful public diplomacy can win a country not just respect but admiration. Examples of the use of soft power include the education of likely future leaders at American universities and publicizing U.S. science and technology, notably the space program, medical advances, and cutting-edge industry. For many years American taxpayers have paid for foreign opinion leaders to visit the United States. President Bush's policies toward Africa and his recent visit to five African countries were probably successful uses of soft power. Many Africans are grateful to the United States for its foreign aid and support for programs to reduce malaria and HIV/AIDS. Soft power, which obviously has much in common with public diplomacy, relies on culture and values to promote goodwill and respect between countries and people.

Public diplomacy is surely about much more than just putting the best spin on government, policies, and leadership. It includes everything the United States can do to improve its reputation. Successful public relations experts always stress that substance matters more than spin or communications. It is hard to get the public to love a company that is known to be a serial polluter, that makes unsafe products, or that treats its employees badly. Indeed, when the truth is disagreeable, public relations efforts alone may be counterproductive.

THE MULTI-FACETED IMAGE

People can feel positively about one element of U.S. policy (e.g. relief for tsunami victims in Indonesia and Sri Lanka) and negatively about others (e.g. the United States' rejection of the Kyoto Treaty or the war in Iraq). Harris polls have shown that an individual can hold very different attitudes to the American president, American policies, and Americans as people. The same person may hold conflicting opinions about the American economy, culture, constitution, political system and judicial systems, and moral and ethical standards.

However, history suggests these different attitudes are linked. When a foreign government implements a new policy, people may dislike the policy, the government, and its leaders but still hold positive views about the country and its people. But that dichotomy does not extend indefinitely. In World War II there were few Americans who believed that, while the policies of Hitler and Japanese Prime Minister General Tojo were awful, the Germans and Japanese were nevertheless good people. How many Arabs differentiate between Israelis and Israeli policies? How many Israelis have positive opinions of Arabs and Muslims, as people? The Iraq War has certainly contributed to negative attitudes toward the U.S. government and its policies, but probably also to the United States as a country and to Americans as people.

American public diplomacy has another handicap. After the collapse of the Soviet Union, there was much talk of a "new world order" and of the United States as the world's only superpower. Before the invasion of Iraq, some American commentators celebrated the fact that they were living in a "unipolar world" and argued that this country was in a position to control, or even dictate, the shape of the new world order, and to bring freedom, democracy, and good government to countries in the Middle East and elsewhere. This talk doubtless fueled fear and suspicion of the United States. Power is seldom associated with popularity.

A further problem is the need for scapegoats. When things are not going well at home, it is convenient to blame others, and powerful countries are easy targets. In the 1970s, 1980s, and 1990s, I was often surprised by the extent of hostility to the United States in Greece and Spain. This was caused, I believe, by the tendency of the Greek and Spanish media and politicians to blame the United States for their economic and foreign policy problems. Rightly or wrongly, Spaniards blamed the United States for abetting the Franco dictatorship, while Greeks blamed Washington for "the colonels," the despotic junta that ran Greece from 1967 to 1974. Many Greeks also blamed the United States for Turkish control of Northern Cyprus. The North Atlantic Treaty Organization (NATO) and the presence of U.S. bases became easy targets for populist politicians in both countries.

In the late eighteenth century, Edmund Burke commented of Great Britain: "I dread our own power and our own ambition; I dread our being too much dreaded.... We may say that we shall not abuse this astonishing and hitherto unheard of power. But every other nation will think we shall abuse it. It is impossible but that sooner or later this state of things must produce a combination against us which may end in our ruin." Thus, as Henry Kissinger notes in *Does America Need a Foreign Policy?*, the challenge facing the United States is "to transform power into consensus so that the international order is based on agreement rather than reluctant acquiescence." *American Exceptionalism*

Americans tend to view the United States as different and special. Many other countries feel the same about themselves; but they often view American exceptionalism very differently. Notably, some of these perceptions were in place long before September 11 or the invasion of Iraq.

In their book *America Against the World*, Andrew Kohut and Bruce Stokes of the Pew Research Center addressed the problem of American exceptionalism. "Nothing is more vexing to foreigners than Americans' belief that America is a shining city on a hill—a place apart where a better way of life exists, one to which all other peoples should aspire." They argue persuasively that "United States citizens are alone in thinking it is a good thing that American customs are spreading around the world." Many foreigners look at U.S. economic and military power, at what the United States says and does, and see not a shining city, not a role model, but hubris and arrogance.

Woodrow Wilson said that God chose the United States "to show the nations of the world how they shall walk in the path of liberty." And Isaiah Berlin wrote that many of Franklin Roosevelt's aides regarded themselves "divinely inspired to save the world." At the risk of making sweeping generalizations, many Americans see this country as the best, the most free, most just, most moral, most democratic, most generous of countries, with the best constitution. That is what American history books tend to teach. Few foreigners see America that way.

They often see this country as having the most powerful military, the strongest economy, and as a land of great opportunity; but many people also see America as money-driven and materialistic, with high levels of crime and drugs. American politicians often applaud (American) "family values." Many foreigners invariably see their own family values as being stronger. Many Americans see this country as caring, compassionate, and idealistic. Many foreigners see exactly the opposite—a rich country indifferent to the poor and disadvantaged, and unwilling to pay more taxes to provide a realistic safety net. Like J. Kenneth Galbraith, they see "public squalor and private affluence." They are puzzled that we are the only Western democracy still to have the death penalty, and that we do not have universal health insurance. While believing in many of the benefits of American democracy, they also see a country where political campaigns require far more money than in any other country, and where half the population does not bother to vote.

THE TRUTH ABOUT FOREIGN AID

There is a widespread tendency in most countries to see their foreign policies as more decent and generous than is the case. In the United States, many surveys show that Americans greatly overestimate how much the government spends on foreign aid, and believe that we are uniquely generous. In one sense we are. The latest available data show the United States providing almost $28 billion dollars in foreign aid, far ahead of Japan ($13 billion), Britain, Germany, and France ($10 billion each).

However, when the data are presented as a percentage of gross domestic product (GDP), the United States ranks twenty-first, spending 0.22 percent of GDP on foreign aid, compared to more than 0.9 percent in Norway and Sweden, and far behind most other European countries which give more than 0.4 percent of GDP. Furthermore, a sizable part of so-called U.S. aid goes to Iraq, Israel, and Egypt for primarily strategic purposes.

THE "SAY-DO PROBLEM"

Complicating matters, is the "say-do problem," in that the U.S. government often seems to say one thing and do another. For example, Washington professes to be a strong supporter of human rights, but the world hears about Abu Ghraib, Guantanamo, "extraordinary rendition," our reluctance to prohibit water-boarding, or refusal to accept that the Geneva Conventions apply to "unlawful enemy combatants." We say we believe in and want to promote democracy, but we support dictatorial governments if we need their support, and oppose democratically elected governments— from Venezuela to Gaza—if we do not like their policies. We have tried to topple unfriendly democracies, and occasionally have succeeded.

Moreover, the United States preaches free trade but provides massive subsidies for agricultural products, imposes legally questionable tariffs to protect American steel companies, and gives substantial price support for U.S. sugar and cotton farmers, freezing out cheaper foreign imports. Washington puts a tariff on Canadian timber imports, in apparent defiance of the North American Free Trade Agreement (NAFTA) and imposes quotas on foreign textiles. These protectionist policies make it difficult for poor Third World countries to compete against subsidized U.S. products in world markets.

In *Rogue Nation*, Clyde Prestowitz identifies many of the reasons why attitudes to the U.S. government have become more hostile. This former corporate executive, who was one of Ronald Reagan's trade negotiators, remarks, "In recent years, America has rejected or weakened several landmark treaties, including the ban on use of landmines, the ban on trade in small arms, the comprehensive test ban treaty, the ABM treaty, the chemical warfare treaty, the biological war treaty, the nonproliferation treaty, the International Criminal Court, and others." Prestowitz also quotes an unnamed British ambassador as saying, "America always preaches the rule of law, but in the end always places itself above the law."

Successful public diplomacy needs to understand the difference between "real" perceptions that can only be addressed by dealing with the substantive issue and misperceptions that may be corrected by better communication. In my experience, public relations people in the corporate world often fail to understand the difference. Public diplomats should not make this mistake.

IT'S THE MEDIA, STUPID

Successful public diplomacy, like successful corporate public relations or political campaigning must start with an understanding of what actually influences public opinion. Of course, events influence attitudes—as do policies and programs—but only as they affect people directly or are reported in the media. The role of the media in reporting events is, of course, overwhelmingly important. Perceptions of leaders, as they are portrayed in the media, are also critical. It is much harder for unpopular leaders to "sell" their policies than popular ones, whether inside their country or abroad. If one does not trust the messenger, one probably distrusts the message.

But public diplomats do not have the option of changing their leaders or governments, and if they cannot influence policy they are left with influencing opinion through the media. Of course, public opinion is also influenced by personal experience and word of mouth, but there is usually little a government can do to influence either in foreign countries. This leaves the media (and not just the news media but, potentially, almost all types of media including comedy, soaps, movies, and more) as a potential tool of influence. Newspapers, television, and radio are much more than mirrors that reflect reality. They are magnifying glasses that can greatly increase or decrease public concerns and shape the agenda of public discourse; they are filters that can give very different views of the same people and events; and they are prisms that can bend opinions.

One reason why American views of the world often diverge from opinions elsewhere is that the media here and abroad report the news differently. News reports about Iraq or the Middle East on American, British, French, and Arab television give widely varying pictures of the same events. Most of them are probably accurate in that they report actual events and show real footage of these events. But the events they choose to report and the video they choose to show are very different. These differences may reflect deliberate biases, but they also reflect the views of editors and reporters as to what is important and what constitutes the "truth." Is it Palestinian rockets killing innocent Israelis or Israeli attacks killing innocent Palestinians? Is it the United States soldiers being killed by Iraqi insurgents or American soldiers killing Iraqis?

If I were unlucky enough to be in charge of public diplomacy I would start with the belief that my goal would be to get more positive, or at least less negative, coverage of the United States and its policies in foreign media. But I would ask myself if this is realistic, or even possible, without changing policies. It is certainly extraordinarily difficult. Of course, public diplomats can help plant some positive stories about the United States in a few media, but influencing the coverage of major events that dominate the news day after day is a huge challenge. The opportunities for American public diplomats to influence the way the world's media report world events are surely very modest.

One difficulty faced by public diplomats is the phenomenon psychologists call "cognitive dissonance," which is the tendency not to accept or believe information that is not consistent with what you already believe. Conversely, there is a human tendency to believe information, even false information, if it supports what you believe. It is also probably true that the stronger your beliefs the more powerful the cognitive dissonance. This surely explains why, five years after 9/11, large numbers of Americans still believed that Iraq possessed weapons of mass destruction, that Saddam Hussein had close links with al Qaeda, and that he helped to plan the 9/11 attacks. It also explains why (as has been widely reported) many Arabs believe that the 9/11 attacks were carried out by the CIA or Israeli intelligence to provide an excuse for America to attack Afghanistan and Iraq. Even if told frequently that this is untrue, many would continue to believe it unless told otherwise by people or media they really trusted.

Ideally, public diplomacy should influence the foreign media, not to present untruths, but to encourage the presentation of truths that are less damaging to our image and reputation. The government and politicians influence the American media all the time, but influencing current events as presented by foreign media to their citizens is much more difficult.

As spin is so difficult, foreign opinion is driven mainly by real world events, as reported by the media we can do so little to influence, and by the perceptions of our leaders. Events are tough to control. In the words of former Secretary of Defense Donald Rumsfeld, "stuff happens"—often nasty, unexpected stuff. Style and rhetoric also make a difference. International criticism of Secretary of Defense Robert Gates is clearly not so strong as it was for Rumsfeld. But, as the U.S. government strives to influence public opinion abroad, public diplomacy should be focused mainly on what the president and administration do and not just how they present themselves and their policies to the world. It may well be true, that as The Economist put it on August 12, 2006, the "Bush administration shows an unmatched ability to put its case in ways that make its friends squirm and its enemies fume with rage." However, a month earlier, the same publication gave public diplomacy a different spin: "Manners and tone of voice matter in international relations...[but] actions speak louder than words." As always, it is likely that the truth lies somewhere in between.

ABOUT THE AUTHOR

Humphrey Taylor is chairman of the Harris Poll, a service of Harris Interactive. He was educated in Britain and has lived in Asia, Africa, South America, Europe, and for the last 30 years, the United States. He has had responsibility for more than 8,000 surveys in more than 80 countries.

From: Humphrey Taylor, "The Not-So-Black Art of Public Diplomacy," *World Policy Journal* 24, no. 4 (Winter 2007/08): 51-59. Used with permission.

Photo courtesy of the family

The LEARN TO LEAD series
is dedicated to the memory of
Lance Corporal **JEFFREY S. HOLMES** USMC,
a former cadet from New Hampshire Wing
who grew from a small, uncertain cadet into
a strong man and a Marine,
all the while keeping his infectiously cheerful attitude.
He was killed in action on Thanksgiving Day, 2004,
during the Battle for Fallujah, Iraq.
He was 20 years old.
Requiescat in pace.